The
STRATFORD CANAL

THE
STRATFORD CANAL

Nick Billingham

TEMPUS

For my daughter, Amber

Acknowledgements

Special thanks for the photographs are due to:
Shakespeare Birthplace Trust Archive
Warwickshire County Records Office
Michael J. Fox
Birmingham Central Library

First published 2002
Copyright © Nick Billingham, 2002

Tempus Publishing Limited
The Mill, Brimscombe Port,
Stroud, Gloucestershire, GL5 2QG
www.tempus-publishing.com

ISBN 0 7524 2122 0

TYPESETTING AND ORIGINATION BY
Tempus Publishing Limited
PRINTED IN GREAT BRITAIN BY
Midway Colour Print, Wiltshire

Contents

Acknowledgements   4

Introduction   7

1.   Conception and Construction   9

2.   Boom Time before the Railways   25

3.   Cuckoo in the Nest, Steam Comes of Age   38

4.   Retirement with God's Wonderful Railway   52

5.   War and Decay   67

6.   Restoration   88

7.   Reincarnation   116

Appendix – Shareholders at the Inaugural Meeting   124

Bibliography   128

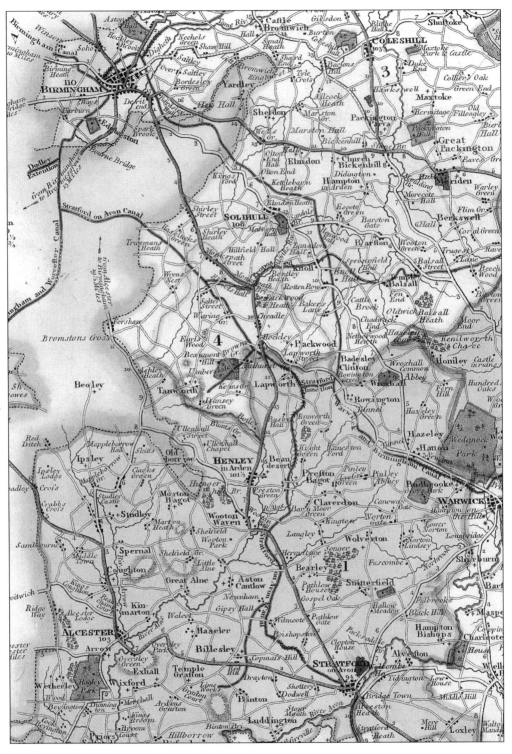

County map of Warwickshire, 1826, showing the Stratford-upon-Canal with its proposed branches.

Introduction

Stratford-upon-Avon may have depended on Shakespeare for its prosperity over the last three centuries, but it has had its share of visionaries who wanted far more than drama – visionaries who started with the canal and went on to change the country. The story of the Stratford Canal embraces the history of transport of the whole nation.

Stratford owes its very existence to the necessity of transport. Its name derives from the place where a minor Roman street crossed the River Avon by way of a ford. The Avon later became a navigation, linked to the industrial Black Country by the canal and, later still, the tramway inspired the age of steam. The canal itself has played a pivotal role in the rescue and revival of the inland waterways system. It may seem to be a fairly modest rural canal, typical of many in the Midlands, but underneath its placid waters there flows a turbulent history.

The canal formed part of the inspiration for the railway network and later, when the railways and roads appeared to threaten the annihilation of the entire canal system, it was the Stratford Canal that pioneered the movement to rescue this vital part of our industrial heritage. As important as the bricks and water are the people who have been associated with the waterway, well known and famous or low paid workers. The waterway has both affected their lives and been affected in its turn.

The story of the Stratford Canal continues into the future – and this work is by no means the end of its history. The canal itself was very much a local venture; its genesis was inspired by Stratfordians keen to better their town, and the money came from people living along its route. As the newly opened canal joined the national network, new ideas and people came to the town, expanding and enhancing it. The tramway, and later the railway system, continued the process but leaving the canal in a kind of limbo, frozen in its Victorian architecture and working practices. This timeless state lasted right up to the Second World War, a small fragment of the early industrial world plodding quietly on despite momentous changes in the nation.

There are not many photographs of the canal in its prime. Photography was in its infancy in the late nineteenth century, when the canal was at its industrial peak, and when it became widespread, this small rural canal was in serious decline. Stratford itself became a well photographed tourist haunt, and luckily a few intrepid photographers took shots of more than Shakespeare's birthplace. Quaint rural life, even the odd horse-drawn boat, found its way into the holiday albums of Victorian ladies. Luckily one such lady managed to catch a glimpse of the last narrow boat to trade on both the Avon and the canal, although she thought she was simply taking a picture of Holy Trinity Church. Such serendipity is rare, but offers a fascinating insight into our history.

Until the canal's restoration the people most intimately involved with it were in no position to afford cameras. Much of the canal's story between the wars is preserved only in the memory of a dwindling number of people. An invaluable resource that has been neglected by traditional historians, and is fast disappearing. Once the canal's rescue was underway photographs become plentiful, offering an insight into the sheer scale of the work involved.

In order to round off the book a few up-to-date photographs are included to show just how the canal has changed in the last couple of decades.

William James.

One

Conception and Construction

Stratford-upon-Avon may be renowned for its flocks of tourists today, but this hasn't always been the case. The town, as we know it, was laid out at the end of the twelfth century. It rapidly became a busy marketplace. From its construction in 1196 until the Civil War it grew in size and prosperity, with the occasional setback caused by plague or being completely burnt to the ground (thatched houses aren't totally idyllic). The town's main income derived from its markets, and these depended on good transport links. The roman road from the garrison town of Alcester to the Fosse Way was an enduring legacy of the past that stood the locals in good stead. The ford, situated between the tramway and Clopton bridges, was difficult to cross when the river flooded, and early on a wooden bridge was built beside it. This wooden affair was pretty rickety and Sir Hugh Clopton paid for the existing stone bridge to be built at the close of the fifteenth century, assuring the town's transport infrastructure. The area to the south of the town was rich agricultural land – the Feldon – producing a surplus of corn since Roman times. To the north was the Forest of Arden equally well off with timber and its by-product, charcoal. Everywhere around, sheep grazed peacefully. As the town bolstered its fortunes, trading and manufacturing blossomed on the back of the wool trade. By the time Shakespeare was born, the town was doing very nicely for itself.

The Avon provided a somewhat informal transport route to compete with the remnants of the Roman road system, although the fights between millers and boaters over water rights left it a haphazard affair. In 1635 William Sandys obtained an Act of Parliament to construct locks and levy tolls between Tewkesbury and Coventry. He probably paid a hefty bribe to the King for this but in the good economic climate it seemed a wise investment. Sandys had watched the building of the locks on the River Thames whilst he was at university in Oxford and saw the benefits that the navigation brought.

The Civil War destroyed his dreams along with the rest of the economy. Sandys was stripped of his rights of the Avon and accused of running a monopoly. William Say was awarded the rights but, as he was one of the regicides, promptly lost them when the monarchy was restored. The Civil War and its aftermath left Stratford, so dependent on external trade, in a trough of economic depression from which it could barely raise itself. The Avon navigation continued, but with little investment. Only Clopton Bridge retained the town's good transport heritage, and that only after the one arch destroyed by the Parliamentarians was repaired. Eventually the majority of the navigation was divided into shares.

In 1664 a syndicate bought several shares in the Upper Avon. One influential member of this syndicate was Andrew Yarranton, who was to publish a book called *England's Improvement By Sea and Land*. He proposed building two manufacturing villages on the banks of the Avon, New Brunswick at Bridgetown and New Harlem where the River Stour joins the Avon at Milcote. He believed that Stratford's position near the heart of England, with both good water and road transport links, would be ideal for such a venture. This was probably the first industrial estate to be planned in its entirety, complete with a transport infrastructure. Sadly, he never managed to raise enough capital to get the project underway. He did however complete a considerable amount of work on the locks of the river. The river was capable of taking boats of up to thirty tons, although it was a difficult journey since there was no towpath, and the boats had to lower their masts for every bridge. The earliest locks built by Sandys were Flash Locks, and they were rather dangerous to operate.

The influential traders of the town realised that something must be done, and not surprisingly they first turned to the Shakespeare Industry. In 1769 the first festival was held, and was a great success, despite being washed out by rain so torrential that the river flooded across the festival site. John Payton, owner of the White Lion pub beside the Bard's birthplace, was in favour of an improvement in trade, as was Mr R.B. Wheeler, an indefatigable author and promoter of the town. Together they had organised the first festival and reaped its rewards. The town's rather anarchic and bucolic atmosphere appealed to many of the London Society that came to the festival, and the tourist trade was born.

Stratford-upon-Avon viewed from the south of Clopton Bridge in 1795. Although the canal had been given its Act, no canal building work had started in the town and the Guild Chapel can be seen across the water meadows that would one day become the canal terminus.

Heading from the Act of Parliament map for the Stratford-upon-Avon Canal.

The latter half of the eighteenth century was one of increasing optimism. The national economy was picking up for a variety of reasons, all of which were reflected on a local level. The Forest of Arden, once a valuable source of timber for construction and charcoal, had been comprehensively felled by the end of the Tudor period. The result was a chronic shortage of charcoal to create iron. Britain was heavily dependent on imported Swedish iron, but Abraham Darby's invention of the blast furnace enabled coal to be substituted for the charcoal. Iron production rose rapidly, and its cost fell accordingly. The invention of steam pumps allowed deeper mines to produce more coal and other ores. The early experiments in crop rotation had also yielded fruit, increasing harvests and thus improving the health of the population. As the industrial revolution began Stratfordians looked about to see if they could take a share in the new wealth of the nation.

The canal system had begun. The Duke of Bridgewater had proved beyond doubt that canals could be both profitable and socially advantageous. The latter merit was deemed of great importance at the time, since the granting of compulsory purchase powers in a canals Act of Parliament, was considered a serious infringement of the traditional rights of the landowners which needed to be balanced by a greater social benefit to everyone. The canal system expanded with James Brindley's dream of a Grand Cross of waterways linking all the major rivers. The canals that served it started to open for business around 1790, and once more proved the great profit to their shareholders.

The River Avon provided the town with transport to the south west, so the influential farmers and gentlemen of Stratford looked to the north and east for their link to the growing

Shottery Brook, to the west of the town centre, was briefly considered as a possible route for the canal to join the river. This beautifully composed picture dates from 1880 and shows Anne Hathaway's cottage in the background. It was a typical working Warwickshire farm.

industrial economy. The Oxford Canal linked the Coventry coalfields to the Thames at Oxford, but the hills between Stratford and Warwick would make a canal difficult to construct despite the advantage of making a link to London possible. The Avon Navigation, although authorised to be made navigable all the way past Warwick had never had locks built on it above Stratford and although there was some coal traffic at Warwick and Barford it never became a through route on a commercial scale. The Birmingham Canal Navigation, straddling both the mineral rich areas of the Black Country and the manufacturing areas of Birmingham, seemed like a good destination, except that the Birmingham Navigation was renowned for being extremely jealous of its trade and water. The local authority originally proposed making a canal to join the Stourbridge Canal in 1775 and sneak into the Birmingham network by the back door, but this scheme never came to anything.

As the web of canals opened, lowering the cost of basic commodities, the pressure for a canal increased. The Warwick & Birmingham Canal wasn't far from Stratford along the River Avon and for a while that looked like a hopeful scheme, but again there wasn't a tremendous urgency to get things done. By 1790 matters became urgent. A through route from Birmingham to London and another from the Severn to the Thames threw the existing coach routes into chaos and left the town virtually isolated. By 1792 a serious proposal to build a canal from Wednesbury coalfields to Stratford arose. It included a branch to Warwick and was intended to carry corn up to the Black Country, returning with coal. At the same time as this proposal was

The river featured in many parts of the local economy, even providing the osiers to make baskets. These were grown in the shallow edges of the river, cut and stripped of their bark like this. This picture was taken in 1880 when the river had ceased to be fully navigable.

the creation of the Worcester & Birmingham Canal. This canal obtained its Act of Parliament in 1791, despite dozens of awkward clauses imposed by the Birmingham Canal.

The Birmingham Canal had become so powerful and so jealous of any rival that it even included a physical barrier between the Worcester and itself, so that all goods would have to be lifted over a six-foot-wide strip of land into another boat. The company charged such high tolls that collieries in the south east of the Black Country around Dudley were being forced to consider building a separate canal to join the Worcester at Selly Oak. This canal made the residents of Stratford and Warwick consider driving their canal to join the Worcester at the same place. The Birmingham Canal was not in the least impressed by this development. It meant that coal from the Dudley and Netherton areas would bypass them en route towards London. A flurry of negotiations behind the scenes finalised the plans.

A canal would be built from Kings Norton, on the Worcester & Birmingham, to Stratford-upon-Avon, another would be built from Digbeth to Warwick, and yet another from Dudley to Selly Oak . The genesis of the Stratford-upon-Avon Canal had happened, but the devil was in the detail. The Birmingham engineer John Snape was commissioned to carry out a survey of the terrain of the route, including two branches to serve quarries at Temple Grafton and Tanworth. The estimated cost was to be £120,000, with a contingency of £60,000.

The Stratford Canal was one of many being proposed at the same time; in 1791, seven canals were authorised with another seven in 1792. 1793 was rightly called the year of canal mania, with a total of twenty new canals, involving nearly three million pounds worth of capital. The Stratford was one of these. On 28 March, Parliament gave it's approval and the subscription list opened.

There weren't many opportunities for profitable investment at the time. The Industrial Revolution was creating wealth, but apart from reinvestment in more industrial ventures, there was little to do with the money. Bank interest rates were low and the new canal schemes provided a much needed outlet for the pent up demand. The subscription list was filled at the

The boats trading on the Avon were modelled on those on the River Severn, but smaller. They had masts that could be lowered to enable them to get under the bridges, and when the winds failed were dragged by hand upstream. The lack of a proper towing path hampered the navigation severely.

The Revd Davenport, like many of the local clergy, invested in the canal and took an active part in its completion. He was extremely energetic in all aspects of town life.

first meeting. John Payton must have rubbed his hands with glee, people crammed into the White Lion for the meeting and no doubt plenty of beer was sold alongside the 1,200 canal shares. The vast majority of the shares were bought by people living close to the canal. Of the 1,200 shares, 776 were subscribed by people who lived within ten miles of the canal's line, the list is a fair indication of the way the wealth of the nation was split. Various lords and baronets were buying shares in blocks of ten, as were all the local clergy. The gentry tended to buy in blocks of five or ten, Whilst the yeoman farmers and widows bought only a couple. Share purchase followed a very different pattern to that of today, this venture was a local project and despite it being the year of canal mania, the shares mostly remained in local hands. Institutional investment such as we have today was unknown, the closest parallel is the share purchases by proxy for the Earl of Plymouth, Duke of Dorset and the 'Mayor and Aldermen and Borough of Stratford-upon-Avon'. Both these latter investments had a direct link to the project. The Duke of Dorset owned thousands of acres through which the canal would run.

Speculation in canal shares was intense, since their value rose sharply once a project was underway, or so it seemed in 1793. Mr James of Henley in Arden bought five shares at this meeting to augment his portfolio of six shares in the Worcester & Birmingham and others. A £100 share could be secured for £5, the balance being paid when the company called for the money. For a wealthy solicitor like Mr James, this seemed to be the perfect investment, as it did to many thousands of investors. His mistake was to then purchase further shares on the open market at a higher cost. For many investors the canal represented a very tangible way in which they could improve their own businesses. Mr William Oldaker bought ten shares. His interest at the time was that he was already using the river navigation to bring corn to his mills near Holy Trinity Church and the canal would mean new markets opening up for him all over the Midlands. For him personally the canal would bring immense success.

The committee was appointed and a sense of euphoria permeates even the minutes taken at the first meeting. With Peter Holford of Wootton Hall in the chair and George Perrott, owner

The first iron aqueduct on the canal posed few difficulties in its construction and encouraged the engineer William Whitmore to build the longest aqueduct in England at Edstone. The first aqueduct over the River Cole at Shirley was a traditional brick arch holding a clay liner for the water. The next at Yarningdale was originally made of wood, although when that finally collapsed it was replaced with an iron structure similar to this. (Warwick CRO, ph350/2604)

of the Avon Navigation, as deputy, the committee comprised of Stratford's best entrepreneurs – such as John Payton and the formidable Revd Davenport, master of Guild Chapel and the Grammar School. A Board of Works was created to oversee the actual construction; they promptly resolved to buy 200 wheelbarrows. At the following meeting they resolved to buy a brand to mark the barrows to reduce theft. Although they had contemplated buying a turf-cutting machine, the construction method would be the traditional army of navvies with spades and barrows. The first section of the canal, from the Worcester & Birmingham at Kings Norton, was to be level as far as Lapworth so there would be no locks to construct, only a tunnel at Brandwood, an aqueduct over the river Cole near Shirley and a fair few bridges. Both the Stratford and Worcester companies wanted their waterways to be wide enough to take barges rather than 7ft 'narrow' boats. Since neither had direct access to the narrow Birmingham system because of the Worcester Bar, it seemed to be worth the effort to make the bridges and tunnels 16ft wide. Nevertheless the Board advertised for its first boats to be built 'as is usually employed', 70ft by 7ft with three-inch-thick oak sides.

The politics of the canal system hadn't finished yet. The committee wanted to extend the line of the canal south east to a junction with the Oxford Canal near Fenny Compton, creating a route to London. This canal would be threatened by a proposed Warwick & Braunston Canal creating a link further north. The Oxford company wanted £5,000 per annum as a guarantee against loss of trade to refuse permission for the Warwick & Braunston link.

The Oxford Canal's high fees caused the plan to be put on the back burner, and while the committee concentrated on building their own line, the Warwick & Napton Canal was authorised, rendering both earlier schemes redundant. The Stratford Canal now negotiated for a link at Lapworth with the Warwick & Birmingham. In 1795 the link was authorised, with a compensation toll to the Warwick Canal. This original plan for a link was to be over a mile long, through a tunnel beneath the village of Lapworth. William Snape surveyed it in the autumn of 1794 and put its cost at £10,000. The powers to raise the additional funds being part of another Act in 1795.

The funding of the canal was going to be an increasing worry for the committee. The Napoleonic Wars had started to bite into England's economy. Where as there had been a surplus of labour when the canal was proposed, now the army had absorbed all the surplus and more and labour costs rose accordingly. Whilst this was happening the Government, needing to finance the war, issued bonds that paid far better returns than the canal company shares, leading to an abrupt withdrawal of further funds for all canal companies. The first bank available for the canal company, indeed the only one in Stratford, had opened in 1790 and was run by Charles Henry Hunt. He sold his business to Messers Horsman and Batersbee in 1796 which was in turn absorbed by the Warwick bank of Whitehead, Weston and Greenaway who traded at 13 Chapel Street.

Hopes for a broad canal, expressed as late as June 1796, faded with the increasing realisation that they didn't have the cash to finish even a narrow one. By August of 1797, squabbles with the Worcester & Birmingham Canal about water led to the construction of the Stop Lock at Kings Norton. The first lock cottage was built beside it at a cost of £150 and a lock-keeper employed for £20 per annum, the princely sum of 38p a week in today's terms.

Shareholders were reluctant to pay the calls on their shares. The situation became critical at the end of 1796, with the canal open only as far as Hockley Heath, there wasn't enough money to pay the navvies and the treasurer had to borrow £100 to avoid a riot. The value of all canal shares plunged, leaving investors like Mr James of Henley in Arden very embarrassed. Luckily his son, William James, had done well enough with his legal practise to bail him out so that he

The Revd Davenport's last sermon in 1838. A new generation was taking over the canal.

Robert Bell Wheler clambered over his fellow passengers to be at the very front of the first boat to pass from the canal to the river. His account of the opening of the canal is both detailed and graphic, although he seems to forget that a workman must have taken a boat through the lock to check that it worked before the opening!

didn't have to forfeit his shares. William had, by the age of twenty-two, become a successful land surveyor as well as a lawyer and even took on some work for the canal company. He was also buying canal company shares as they became available – at rock bottom prices.

The construction proceeded in fits and starts as the funds became available. The massive flight of locks at Lapworth, and indeed the rest of the canal, had to be reduced in scale to the 7ft width of the other Midland canals. The ideal of a barge canal had foundered, indeed the Worcester & Birmingham Canal, also in difficulties, abandoned their plans for a wide waterway at the same time. In 1799 the company obtained a third Act of Parliament to move the line closer to the Warwick Canal. Samuel Porter was appointed Engineer and set to work on the locks and link.

On 24 May 1802, the canal finally joined the Warwick & Birmingham. It was still thirteen miles short of Stratford, but at long last it was a through route which could earn some money. The workmen were treated to £3 3s worth of drink to celebrate. Quite what the residents of Lapworth thought of this is fortunately unrecorded. Coal and other goods began to move along the new route, although the tolls for a long journey were substantial and the Stratford company had to reduce theirs to encourage traders to use their route. The committee became embroiled in a series of disputes with other companies about just how much could be charged for journeys across several canals. An uneasy balance occurred, a mixture of tolls and discounts for various routes, making budgeting a nightmare. The balance would occasionally be rocked by a proposal for brand new canals to bypass existing ones. The Central Junction, from the Stratford Canal to Abingdon, was one, and the 'Union Canal', from the Stratford at Earlswood to Stretton on the Oxford, another; both were advanced and researched but never built because their intended victims (the Oxford and the Warwick, respectively) dropped their tolls.

As far as Stratford town was concerned, nothing was happening at all. The Corporation wrote to the company asking them to hurry up in 1803, but the reply was that there weren't the funds to go faster. In 1808 they even offered to lend the company £2,000 to get a move on. There was an air of apathy over the project and little more digging took place.

What the company needed was someone with the drive and determination to overcome all the petty obstacles and complete the project. William James, by now a very successful land agent, lawyer and entrepreneur, rejoined the committee in 1808 and gave the necessary push. He had contacts all over the country, and just as many business interests. He now had a substantial holding of the companies shares, so it was definitely in his interests for the project to be completed. He joined the Board of Works in December 1810.

With renewed vigour the digging restarted and new plans laid. In 1810 a new share issue raised more funds and in the background more ambitious plans were afoot. The original plan of the canal did not have a link with the navigable River Avon in Stratford. This would have severely upset the Worcester & Birmingham Canal whose goodwill was so vital in the early years, as it would have provided a competitor for the rich Birmingham to Bristol trade. William James didn't place credence in goodwill for the Worcester & Birmingham, his own father had been reduced to poverty by speculating in their shares. A link to the Avon there would have to be. Various options were considered and the final route continued the existing canal from its terminus at One Elm Basin down the side of the town's rubbish dump to a large expanse of water meadow called Bancroft beside the Avon. By now Deputy Chairman, William James was determined this plan would succeed. William Oldaker was fully in support of this; the Stratford Mills had become extremely profitable using the Avon Navigation, so much so that William had become an equal partner with John Tomes and William Chattaway to form the Stratford-upon-Avon Bank situated in High Street.

Alongside the new route were variations to the canal's architecture. With such a tight budget any bridge had to be built as cheaply as possible. William James was familiar with the use of cast iron, he owned a foundry at Birchills, and a new design of farm accommodation bridge – the split bridge – was put into effect. Probably copied from those on other canals, this bridge had cast iron decks supported from each side, leaving a gap in the middle for the boat's tow rope to

With the canal open the river needed major improvements to allow narrow boats to navigate downstream to the Severn. This is the old mill bridge below Holy Trinity Church being widened in 1867.

pass through. With a nominal width of 7ft 6in at the waterline, and no towpath, it was much cheaper than a brick bridge, and could be prefabricated in bulk. The proposed staffing of the canal was left vague, lock cottages being built only as far as Preston Bagot, and these were two-roomed hovels built around bridge formers, now the famous Barrel Roof cottages.

Aqueducts were essential for the remaining length, the first at Yarningdale Common needed to be 30ft long and was easily constructed of wood. The next two, at Wootton Wawen and at Edstone, were another matter altogether. Both would be incredibly expensive in brick and clay. A radical approach to their material was needed.

As midsummer 1813 dawned, William James had been placed completely in charge of the project. He had convinced the committee to invest in a light gauge tramway to speed up the digging; with the result that the canal reached Wootton Wawen on the evening of 22 June. They immediately started transhipping coal from boats to wagons to serve Stratford and Alcester. The canal terminated in a basin just before the site of the aqueduct. In September the plans for the route of the canal to link to the river were deposited, and also, twenty years late, the all important reservoirs at Earlswood, to ensure a stable water supply. The weather had been pretty wet at the close of the eighteenth century, 1,079mm of rain fell in 1799 in central England, but this dropped markedly in the first decade of the nineteenth, only 734mm in 1805 and 828mm in 1813. The engineers realised that a reservoir was essential to maintain the levels in the canal between wet and dry years.

William James further incensed the Worcester & Birmingham by buying all the shares in the Upper Avon Navigation. A race was now on to open the first route between Birmingham and Bristol. Whoever won would be wealthy, and William James intended that it should be him.

The last hurdle was the two aqueducts and a lock flight down into the town. The resident engineer was now William Whitmore, who no doubt had a few sleepless nights over the new design of aqueduct – a cast iron trough supported on brick piers. The concept had been pioneered at Longdon on Tern, but the radical difference here was the prefabricated sections.

The committee must have had sleepless nights too, at every meeting anxious concerns were raised. Wootton Wawen Aqueduct was built easily enough, apart from the local landowners protestations that it was an eyesore (Lady Smythe of Austy Manor sued the company for the return of the land she had sold them, but with William James acting for the company, she lost the case after several years and paid out thousands in costs). Edstone Aqueduct was an altogether more ambitious undertaking as the longest aqueduct in England, but again there were no special problems and the canal started the last few miles to Stratford. William James suggested that the canal from Wootton Wawen should be replaced with a tramway, which would be easier and cheaper to build, but nothing came of that suggestion.

Geology at the start of the nineteenth century was something of an art rather than a science. The quarries of limestone at Temple Grafton were going to be served by a branch canal. The original planners, however, did not realise that there was a ridge of blue lias limestone reaching from Edstone Aqueduct, past Temple Grafton to Bidford. Once the actual digging started it became obvious just how much stone there was. William James negotiated the purchase of land around Wilmcote far in excess of the needs of the canal. By paying for the land himself, and then selling a bit to the canal company he found himself the owner of yet another quarry. The limestone here was in two types of strata, one suitable for building stone, the other just right for firing in kilns to make quicklime.

The Old Stratford flight of locks posed no special problems, although they were built as cheaply as possible. The only hold-up was the weather, January 1814 was the coldest for nineteen years, with an average temperature of -2.9°C. Unfortunately for the company the race with the Worcester & Birmingham was lost as the last few locks were being built. The Worcester opened in December 1815.

The Stratford-upon-Avon Canal was finally opened on 24 June 1816. A grand procession of boats swept into the terminal basin and one, an inspection boat on loan from the Worcester & Birmingham Canal, passed through the lock onto the Avon amidst the cheers of thousands of onlookers. Mr R.B. Wheler was on it, clambering over his fellow passengers to get to the bows and to be the first person in the first boat to unite canal and river.

It had taken twenty-three years and cost £300,000.

At first, lift bridges were used for farm bridges. This one is at Hockley Heath; later on, split bridges were used.

The summit level of both the Stratford and the Worcester Canals were built for wide beam barges. Local folklore has it that a wide beam boat was used on the northern section of the Stratford. However this boat is actually on the Worcester & Birmingham Canal at Primrose Hill. Possibly another boat traded between Kings Norton Stop Lock and Hockley Heath.

Barges could use this route until the narrow guillotine stop lock was installed beneath Lifford Lane. The gate could operate regardless of the water level in either canal and helped both companies preserve their vital water supplies.

Extract from the Act of Parliament map for the Stratford-upon-Avon Canal.

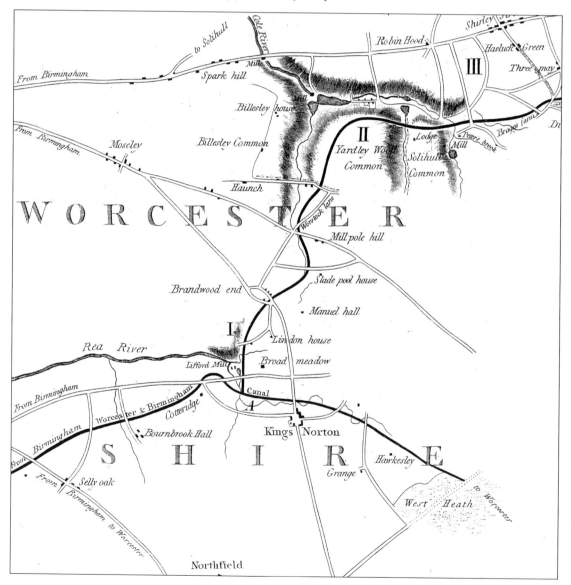

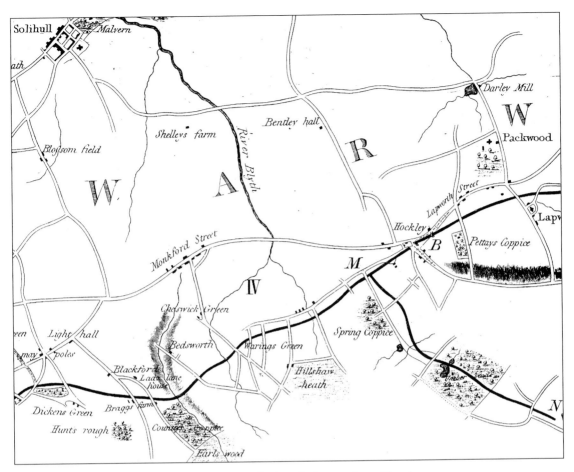

Extract from the Act of Parliament map for the Stratford-upon-Avon Canal.

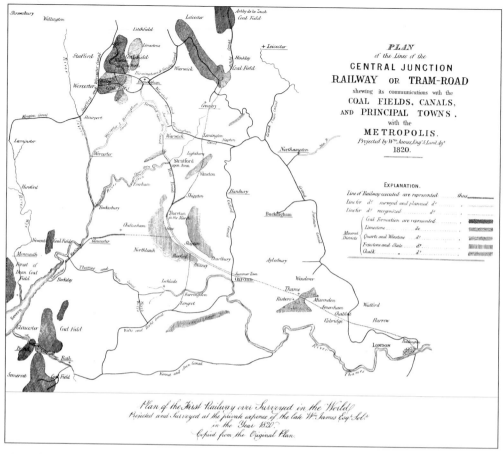

Plan of Central Junction Railway as conceived by William James. Almost as soon as the canal was open, William James was planning the world's first railway. The route he proposed was from Stratford to Paddington in London, using steam locomotives. The scheme was stillborn although the first section to Moreton-in-the-Marsh was built as a horse-drawn tramway and opened in 1826.

Two

Boom Time before the Railways

The festivities of the canal's opening soon gave way to serious trading. The wharf at One Elm to the north of the town was serving half a dozen lime kilns and a brick works. The basin on the Bancroft could only handle nine or ten boats unloading at once, with a few others descending through the lock to unload off the river. The original plan was for some boats to carry on downstream to take coal to Evesham, but the river locks were too small to take the normal Midland-type narrow boats and the coal had to be transhipped into river boats. William James had to start investing heavily to improve the river navigation.

The traffic congestion naturally caused frayed tempers and right from the beginning there were altercations between the boat crews and canal staff about the use of locks, and in one case the canal company had to take a boatman to court to make him pay damages. There were some very harsh penalties available to the canal company. If you sank your boat and blocked the navigation, the fine was 5s an hour and all the costs of re-floating the boat. Wasting water could incur a fine of £5, as would overloading.

Thomas Brewin, by now the company's main contractor, suggested that another six lock-keepers be appointed, with cottages set at strategic intervals between Wootton Wawen and Stratford. These to include a carpenters yard for the constant repairs that were needed. The 1815 Act specifying the Earlswood Reservoirs had to be implemented and the link to the Warwick & Birmingham Canal needed to be sorted out. The Warwick Company was not in the least convinced that it was getting adequate water from the 1812 lock now that the boats were continuing past the link down to Stratford.

The tramway that William James had used to complete the canal was now redundant, and the company sold it to his quarry at Wilmcote. The branch canal to Temple Grafton remained on the wish list, the canal widened out at the junction, but since the route started directly into a deep cutting it would prove costly to build, and the tramway in the quarry could obviously be extended along the route at far less cost. William James abandoned his position on the Board of Works, partly amid an investigation into his somewhat anarchic business methods, and partly because of his other commitments. He did remain an active committee member.

The stop lock built in 1812 to join the Warwick and Stratford Canals, and filled in shortly afterwards. Now rebuilt by British Waterways. Once the canal opened to Stratford, the Warwick & Birmingham Canal Company insisted that this lock was replaced by one that gave them a full lock full of water.

Although the canal was now open and earning money, the new works, including a second basin on the Bancroft, meant that there was precious little left over for the shareholders. The company was saddled with a huge debt which soaked up most of the profit. Earlswood Reservoirs were completed and the guillotine-gated lock was replaced with one with a normal fall above lock 21. Another unlooked-for expense was the replacement of the gates in the main Lapworth flight, unchanged since 1799. The canal offices also had to be moved to Stratford.

Amidst all this, boats were pouring down to Stratford with coal and agricultural machines, returning laden with wheat and barley. Limestone from Wilmcote quarries was now cheaply available in Stratford and the town started a rapid expansion with housing estates appearing on its periphery. Although the shareholders were seeing little return, the town and outlying districts were becoming steadily more prosperous. John Payton,s son, also named John, built several new streets on the land enclosed between the town and canal; John Street, Payton Street and Tyler Street are all direct evidence of the prosperity the canal brought to the town. Mind you, the tourist trade was booming, as Shakespeare, Mulberry and Great William Streets show. These were also built by John Payton, who finally sold off the White Lion to help fund the project.

Even as the canal's trade blossomed, a new technology was being born right on its doorstep; a technology that would eventually destroy their commercial viability. When William James realised how cheaply and simply the tramway in the Wilmcote quarry could be laid, it started him thinking of a far more grandiose scheme – a railway. He wasn't alone in this. Edward Pease in Darlington was thinking along the same lines.

The weather was a critical factor in canal transport. Too little rain meant that the boats slid to a halt on the muddy bottom. The year of the canal's opening was seriously wet with 1,007mm of rain, so that there was little competition between the companies for the water, but 1817 was dry, only 882mm. All through the long summer the committee kept a close eye on the water levels, congratulating themselves that they had had the foresight to dig the summit length extra deep to act as a reservoir, nevertheless other canals in the system were closing for lack of water and the overall trade was severely restricted. Winter's ice also blocked the navigation and damaged the canal's structures. A constant watch was kept on the more novel features like the cast iron aqueducts, fortunately they survived although the cheap brickwork of the piers took a battering each winter. The first few winters weren't too bad, but in January 1820 the average temperature was -0.3°C and traffic became locked solid, it happened again in 1823.

The canal's operating system soon established a routine. Events beyond were already starting to cast long shadows however. By 1819 William James had conceived a Grand Junction Railroad in his private notebooks. Realising, as had Andrew Yarranton before him, that Stratford was ideally placed at the heart of England's trade routes, a railway from Stratford to London, with a few spurs, would create wealth beyond the dreams of avarice. Particularly for some one who owned the Upper Avon Navigation and shares in the Stratford Canal like himself.

Wealth creation was the driving force behind the construction of the canal, but in the twenty-five years since the project had started, England's economy had changed beyond all recognition. The Napoleonic Wars had caused the government to suspend the convertibility of bank notes, realising huge amounts of credit to fuel the war effort. With the war over, the time of reckoning had finally come. The gold standard was reintroduced and paper money could once again be exchanged for cash. To everyone's horror it transpired that nearly half the notes in circulation were either forged, or drawn on provincial banks that had no reserves. Mr Oldaker's Stratford-upon-Avon Bank was one of the few provincial banks not to be compromised in the resulting shake up, indeed its story is one of continuous trading right up to the present day through various mergers and amalgamations (it has branches in Hong Kong and Shanghai today).

The economy ground to a halt between 1818 and 1820. The practical effect was to force entrepreneurs and industrialists to look for ways to reduce their costs, transport costs in particular. One entrepreneur particularly badly hit was William James, who had most of his wealth tied up in canal shares and land. It gave him a serious cash flow problem.

The Grand Union Canal at Lapworth in 1910. The boat here is one of the type that collected the tar from Stratford Gas Works on a regular basis right up to the end of it's working life. The tar was an obnoxious cargo, although a valuable source of chemicals. The railways were happy to leave this sort of freight on the canals. (Warwick CRO, ph108/07)

These buildings were the original canal offices, subsequently rented out to local businesses as the trade on the canal declined and the canal's management took place in the Great Western Railways offices.

The immediate impact of this on the Stratford Canal was William James' scheme for the construction of a tramline from the basins at Stratford to Moreton-in-the-Marsh. It was the first part of his scheme to reach London, although the shareholders never knew this. Lord Redesdale 'the principal backer' was aware of the plans, although he viewed them with some scepticism. The venture could provide nothing but benefit to the canal, opening up new markets in an area hitherto inaccessible, and delivering their produce right to the canal wharves. Many of the shareholders of the canal, such as Thomas Brewin, the main contractor, and Richard Greaves, one of the local hauliers, invested in the project.

The canal had to give generous allowances to carriers using the whole route, which reduced its profitability. Companies as large as Pickfords included the town on their list of regular destinations, and smaller companies soon flourished, transporting parcels and people far and wide. Passenger transport was a lucrative trade from Stratford as the opening of the canal system had thrown the existing network of coaches into confusion. The tourist trade was getting underway in the town and a project to build a theatre seemed to have good prospects. Until now travelling theatre companies had to perform in tents, barns or the town hall. Shakespeare's birthplace attracted all sorts of visitors, some hoping for inspiration, some simply to admire the quaint, bucolic little townhouse.

The existing canal was serving the needs of the town well, but the parlous state of the River Avon Navigation was restricting movement of goods to the south west. William James was investing heavily in repairing the locks, as well as investing huge amounts of his money in the tramway project. When these two schemes were complete, trade on the canal would improve and earn him his well-deserved reward, his other schemes involved equal expenditure. The Liverpool & Manchester Railway was one such, and there were another dozen railways in which he was involved. He suffered a major cash flow crisis in 1823, which left him in Fleet Street Prison for a while, and then he was declared bankrupt. His shares in the Avon Navigation went up for sale and were bought by a consortium of local carriers and industrialists. His active participation in the affairs of the canal and tramway was effectively at an end; although his legacy would echo down the corridors of time.

The tramway opened in 1826 amidst as much celebration as the canal a decade earlier. The canal immediately saw a massive increase in trade as coal poured down to the untapped markets to the south, and barley and wheat surged north to the ever hungry masses of the Black Country. Richard Greaves and his fellow proprietors of the Upper Avon finally improved the locks enough to encourage trade down the river to Evesham and beyond, invading territory previously supplied by the Worcester & Birmingham Canal. The coal trade alone was in excess of 10,000 tons a year, and the Vale of Evesham was already renowned for its vegetable exports.

All around the town the booming trade generated by the canal was changing peoples lives. John Tallis, a lock-keeper, bought a plot of land on the Warwick Road beside the canal bridge and built a house on it. He had seen that there was a new business opportunity in providing stabling for the horses that pulled the boats, and another to quench the thirst of the boatmen. The house opened as the Warwick Tavern in October 1831, complete with a snug and a parlour, rooms for guests to stay overnight and, as each year passed, more and more stables. Next door to the tavern was a small boat builders yard, complete with its own dry dock.

The basins in Stratford were working at the maximum capacity in the 1840s. There were two dry docks for boat building and repairs, a handful of coal merchants and timber wharves. The tramway was extended with sidings, turntables, swing bridges and weighing machines to ensure a speedy transfer of cargo. The canal company was at long last making some money with a gross tonnage of around 150,000 per annum. It had survived a variety of disasters including the failure of the wooden aqueduct at Yarningdale. Too much rain had weakened an embankment on the Warwick & Birmingham Canal to the point that it burst. The resulting torrent swept down the little brook on 28 July 1834, sweeping the flimsy aqueduct away. The replacement was built to the same design as the original iron troughs at Wootton and Edstone.

Coal was the staple commodity of the canal system, and Stratford Gas Works had its own unloading basin just north of One Elm Lock (Lock 52). This photograph was taken around 1910 when the gas works had moved 200 yards further up the canal to be closer to the railway freight sidings. Although the gas works had its coal delivered by rail at this point, there was still the tar oil by-product to be shipped out by boat.

The town's industry, apart from the Shakespeare trade, was advancing with these new transport links. Richard Greaves expanded the quarries at Wilmcote to the point where they were producing 16,000 tons of limestone a year – all transported by canal. He had also introduced a new type of cement in which lime was clinkered with clay to produce a product very similar to modern cement. This was exceptionally popular and was sold as far afield as London. Wilmcote quarries even produced the flooring stone for the House of Lords when it was rebuilt. In the town itself an iron foundry started, as well as brick works and a gas works, all situated around the small basin at One Elm. Since Stratford was so close to the barley-growing areas of the Feldon, it was only natural that a brewery should flourish, and Edward Fordham Flower soon saw his opportunity. He started his business in partnership with James Cox, the timber merchant, at a wharf near One Elm. Soon he was expanding and had built a large-scale brewery. James Cox continued with his timber business, moving his wharf to the basin. Although the partnership dissolved, Cox's supplied all the cooperage for the brewery. Flowers were also supplied by another company, Kendal's. They made the glucose-based brewing syrup, the ingredients for which were brought to the town by various companies and private boat firms. Many of these industries were run by Non-Conformists, and a proportion of their profits were used to improve the lot of their labourers. Better housing, churches and schools all started to appear along the banks of the canal close to One Elm wharf. The wealth the canal was creating was at last starting to improve the lot of ordinary townspeople. It was a slow process though, throughout the 1840s and 1850s the growth of the town appears slow because of the recession that occurred when the gold standard was re-introduced.

Flowers Brewery built their first premises beside the canal. It became one of the principal industries of the town and continued brewing until taken over by Whitbreads in the late 1960s. As Stratford was close to the grain growing district of the Feldon, to the south, the canal opened the way to transport the beer further than a solely local market. As Flowers prospered so to did its suppliers such as Kendals, who made a brewing syrup, and Cox's timber merchants, who made the barrels.

The canal encouraged other new businesses. John Tallis, an enterprising lock-keeper, built this as the Warwick Tavern in 1831, to serve the needs of the boatmen and their horses. Other service industries included three boat builders and several other pubs.

The original planners of the canal recognised that limestone would be a valuable cargo to the blast furnaces of the Black Country. Once the canal opened, the quarry at Wilmcote could export its stone. The village expanded suddenly with twenty-eight new cottages built in the 1840s.

The earliest surviving photograph of Wilmcote flight. In 1892 the canal was in tolerably good condition and maintained by the Great Western Railway. Although it is a popular belief that the railways deliberately killed off the canal, the evidence of the brickwork shows that they actually carried out a lot of repairs on the locks and bridges. Although the towpath and lock seem overgrown, the lock island on the right has been dug over. The lock-keeper grew his potatoes here.

Opposite: *Extracts from the Act of Parliament map for the Stratford-upon-Avon Canal.*

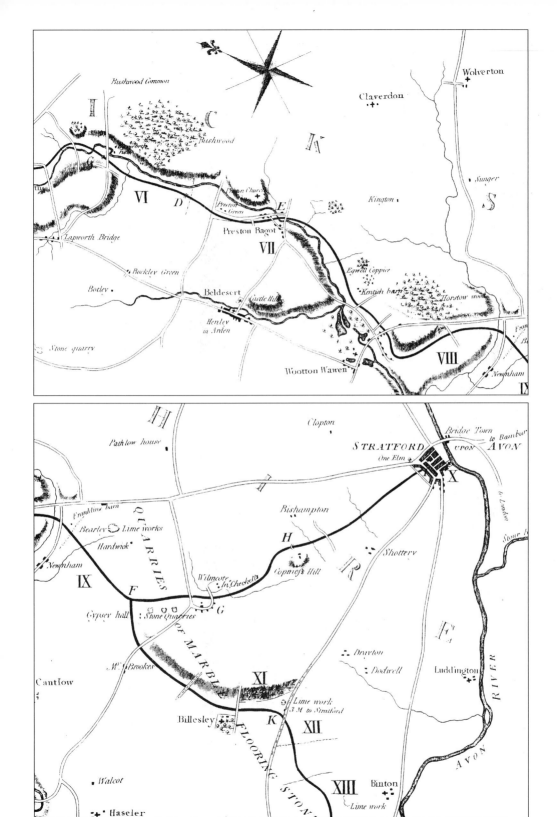

33

A BIT OF THE BLACK COUNTRY (IV)

Gradually the boatmen moved onto their boats as trade declined in the 1870s.

Here a pair of F.M.C. boats are tied up. This company was one of the few that looked after their staff well.

Edstone Aqueduct proved to be an enduring structure. The longest aqueduct in England and, at the time of its construction, a miracle of modern engineering. The cast iron trough has been reputed to be designed by Marc Brunel, although there is no evidence for this, a more likely designer would be William James, who just happened to be Chairman of the Board of Works, as well as owning an iron foundry at Birchills.

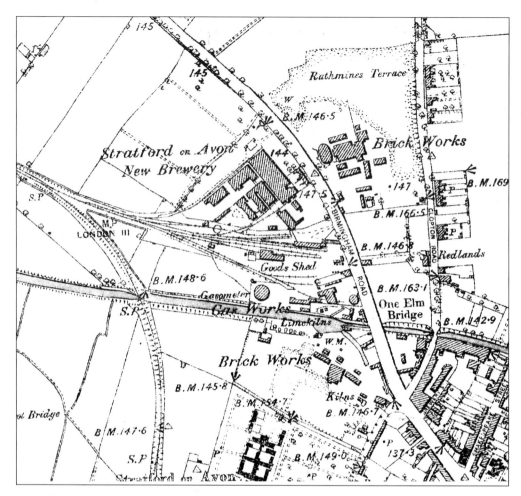

One Elm Basin and the Stratford-upon-Avon Railway's first station. One Elm was designed as the original terminus for both the canal and the Stratford-upon-Avon Railway; both ventures carried on from here to link up with other routes. This map dates from 1885 and already the arm of the canal south from One Elm Lock has already been shortened by about 200ft as fewer businesses traded on the canal. The brickworks and limekilns have all vanished beneath post-war development. The railway line crossing the canal on the left of the map is the link between the Stratford-upon-Avon Railway and the Oxford, Worcester & Wolverhampton Railway to the south of the town. As both these railways were built as mixed gauge (7ft and 4ft 8in), this 4ft 8in link represents the end of Brunel's dream of a wide gauge railway system.

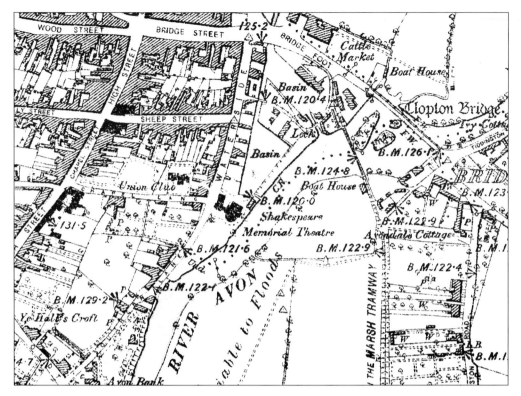

The canal basins on the Bancroft. In its heyday the Bancroft was a thriving inland port. This 1885 survey already shows a decline however, with approximately 30% of the tramway tracks having been lifted. The Board of Health survey of 1855 shows the full extent of the tramway interchange, unfortunately it is on such a massive scale that it can't be reproduced clearly here.

Three

Cuckoo in the Nest, Steam Comes of Age

In 1845 William James' legacy became apparent. During the planning of the Stratford & Moreton Tramway, he had specified that the motive power was to be steam-driven locomotives, specifically an engine called the *Invention*, at a cost of £400. His idea had to be abandoned partly because the gradients on the line were too steep and partly due to the other directors innate suspicion of such a revolutionary concept. Never the less Mr James persisted with the idea in other areas in partnership with a young engineer by the name of George Stephenson who he introduced to Edward Pease. Pease was then building a horse-drawn mineral line between Stockton and Darlington. James was also surveying the line of a railway between Liverpool and Manchester. Today these lines are considered the first of the railway age. The Stockton & Darlington, opening in 1826 and the Liverpool & Manchester in 1830, revolutionised freight transport. The technology of a canal system is very much that of a pre-industrial society, the railways required a large and accurate industrial base from which to draw their expertise. The canal system had created the climate for the industrial base and from that came the railways.

As the Stratford Canal Company thrived during the 1830s and 1940s the railway era started to gather momentum. In the north of England companies were being formed to build railways to a gauge of 4ft 8in whilst in the south the Great Western and her allies were seeking to build to a 7ft gauge, realising the ambitious plans of that bold engineer Isambard Kingdom Brunel. This Battle of the Gauges was to have a profound effect on the story of the Stratford Canal.

In 1842 the canal company leased the shares of the Upper Avon Navigation in order to secure the route. However by 1845 railways were enjoying a huge boom and tentative schemes were being promoted all over the country. It was like 1793 all over again, but being played for higher stakes. The politics between the rival gauges, and even individual companies is a tangled web.

Stratford once again found itself at the centre of attention since it was situated between the broad gauge in the south, and narrow gauge in the north. The canal company realised that there was no way that it could compete against the railways for freight traffic. The era of canal transport was effectively doomed when the Stockton & Darlington opened in 1826. The directors of all the canal companies were faced with the problem of how to salvage something from their investment. Ownership by a railway company was a fairly safe option, partly because the natural topography of the land meant that the ideal route for both rail and canal was more or less in the same place – in some cases exactly the same place, as well as being between the same termini.

The empires of broad and narrow gauge railways constantly sought inroads into each others territory. The Great Western Railway advised and backed schemes that would extend its sphere of influence northwards; these were the Birmingham & Oxford Junction Railway and the Oxford, Worcester & Wolverhampton Railway in the region affecting the Stratford Canal. The OWWR, the Old Worse and Worse as it became nicknamed, received a guarantee of £4 million from the GWR for its scheme to build a broad gauge line from Oxford, through Worcester, to Wolverhampton. The directors of the OWWR failed to mention the limit to their shareholders, and went on a spending spree, buying up not only land relevant to the construction of their main line, but other concerns that might possibly come in handy. Two of these were the Stratford & Moreton Tramway and the Stratford-upon-Avon Canal. The strategy behind this was to build a branch from their main line at Moreton-in-the-Marsh to the south of Birmingham. The lease of the tramway was agreed and the sale of the canal as well. They would take place when the company obtained its Act of Parliament. The Birmingham & Oxford Junction Railway started off as the GWR's back stop. If the OWWR failed to get its Act then they would have powers to purchase the canal, although their main plan was to build a broad gauge branch from their main line near Warwick into the town.

The wrangling amongst all these companies was intense, with the GWR by far and away the most experienced. The Great Western was rapidly losing faith in the board of the OWWR, partly since the money they were spending was out of all proportion to the value of their line, and worse, well over the guarantee. Once the OWWR started to get cold feet about building a broad gauge line, the Great Western backed away as fast as they could.

The Parliamentary session of 1845 must have been mayhem. There were dozens of petitions for railways from all over the country. At the start of the year the OWWR and the Birmingham & Oxford Junction agreed between themselves that both would seek powers to purchase the

Mr Flowers' brewery was a major business in Stratford, requiring modern transport.

Wilmcote Quarries used the canal to transport over 16,000 tons of stone and lime a year at its peak production. The company had its boats built and registered in Stratford. The advent of the canal allowed the small hamlet of Wilmcote to grow into a substantial village. The quarries employed several hundred people.

canal, that the two concerns would link at Stratford, providing a new north east to south west route. This new route would mean that the canal need not be converted to a railway because a track would be formed into Birmingham via the Birmingham & Oxford Junction branch. Stratford would become a thriving interchange itself, with the tramway line from Moreton-in-the-Marsh meeting the OWWR branch from Honeybourne and the branch of the Birmingham & Oxford Junction. There would be an immediate problem if the Old Worse and Worse announced its plan to convert the canal to a railway since this would compete for traffic with the Birmingham & Oxford Junction. Another factor against this conversion was that the work would necessarily entail the closure of the canal whilst the railway was constructed, leaving Wilmcote quarries without any way to ship its stone to market. These quarries were producing huge quantities of stone, lime and cement and Mr Greaves, the owner, had no intention of letting his business be strangled by this.

On 27 July 1846 the Oxford, Worcester & Wolverhampton Railway Co. was granted its Act of Parliament, including the power to buy the Stratford Canal as it had agreed. The terms mentioned nothing about converting it to a railway, only to maintain it as a canal. As the company thought that the actual purchase would then be completed and paid for by the Birmingham & Oxford Junction Railway Co. they had neglected to add a clause enabling them to raise the required money to buy the shares. The Birmingham & Oxford failed to get their Act passed, leaving the Old Worse and Worse in a tight corner.

The railway company had to defer the completion of the purchase of the canal, paying interest on the capital to the canal company. They hoped that the Birmingham company would get their Act of Parliament quickly, and thus want to buy the canal off them, but this simply didn't happen. The Great Western Railway, always a prime mover behind the scenes, was negotiating with Mr Greaves to build an entirely new branch to Stratford that didn't require any part of the canal to be closed. This would become the nominally independent Stratford-upon-Avon Railway. This plan left the OWWR with a canal it couldn't sell, but was bound to maintain in working order. It still failed to purchase the concern outright but paid interest instead. This uneasy situation persisted whilst the GWR sought new alliances.

By 1849 the purchase was still not complete. The OWWR had spent a huge amount of money and still not started its own main line, its only income was from the tramway and a few other small concerns. The purchase date was renegotiated to 1856 to allow the company to finish its line and earn some money. Gradually the line grew from Oxford, first to Moreton where it could link to the tramway, and then as far as Evesham. Each extension reducing the freight carried on the canal. The canal revenues started their inexorable slide downwards.

The railway company managed to get an Act of Parliament enabling it to buy the canal shares in exchange for rent charges. This helped spread the cost of the purchase over some years after the official date. On 1 January 1856, the canal company was at long last officially purchased by the OWWR.

The next four years were to see a small increase in traffic, despite the steady trend downwards. The Birmingham & Oxford Junction Railway had fallen by the wayside, but the Stratford-upon-Avon Railway had taken up the idea and, with Mr Greaves as chairman, started construction of a mixed gauge line from Hatton to Stratford, with a link to the quarries at Wilmcote. The mixed gauge line was the compromise solution to the two different gauges, it had three rails so that trains of either gauge could use it. The Stratford-upon-Avon Railway and the Oxford, Worcester & Wolverhampton Railway built their lines to the town like this.

In 1860 the railways converging on Stratford from the north and south were completed and then linked in 1861. Both the Stratford Railway and the Honeybourne branch of the OWWR were mixed gauge, but they were linked by a stretch of track built only to 4ft 8in. The battle of

Bridgefoot in the heart of Stratford found its trade gradually disappearing as the railways grew. Today only the boundary walls and Cox's Yard remain to remind us that Stratford hasn't always been so reliant on tourism. This site decayed as the canal declined, and was converted into a bus station before being incorporated into the gardens of the Red Lion pub.

The industrial area beside Cox's yard and the tramway terminus became redundant. The tram rails were lifted after the First World War and the buildings found alternative uses.

the gauges was over. The resulting rail network was disastrous for the canal. Passengers and freight rapidly moved onto the railways. The new owners had financial problems in profusion, and the OWWR and several other small railways amalgamated to form the West Midland Railway, but this wasn't enough to give them the necessary strength, and in 1863 they were subsumed into the Great Western Railway system.

The canal ceased to be an individual company once purchased by the OWWR. The directors and committee members cast off their mantle of responsibility, no doubt with a huge sigh of relief. The canal had never shown the kind of profit that was expected of it. Despite its early promise it had never formed part of a through route between industrial centres and as such had only been used for agricultural produce and the kind of light industrial goods that Stratford manufactured. The northern part of the canal, from its junction with the Warwick & Birmingham to Kings Norton, saw much greater use as a short cut on the route between London and the Black Country, but since this section faced direct competition from other canals, the revenue raised in tolls was always low, barely enough to cover the maintenance costs.

From 1857 the waterway was part of a much greater concern, whatever the name was. Routine maintenance became the job of a canal manager, under instructions to spend as little as possible. Staffing levels were reduced, and wherever possible freight was encouraged to move onto the railway.

Not all freight vanished from the boats. The GWR didn't particularly want the job of moving sewage and rubbish out of Birmingham on their nice clean trains, so some regular traffic remained on the northern section. Neither did any of the railway companies want to move the tar oil waste resulting from cooking coal in gas works. The Stratford Gas Co. was producing about twenty tons a month, and this went off in a boat. The oil contained a brew of chemicals which although noxious, gave birth to the aniline dye industry, bring new and vivid colours to the previously dull clothes of the Victorian era.

Wilmcote Quarries, by now called Greaves, Bull & Lakin, continued to use their extensive fleet of narrow boats between Wilmcote and newly opened works at Stockton on the Grand Junction Canal, and from there down to London or to their wharf at Gas Street in

Birmingham. The tolls on the canal system were dropping rapidly as the canal companies sought to maintain their market share. The boatmen too were working for lower wages, and so for a company with an existing fleet, canal transport remained a moderately attractive transport method. Nevertheless, Greaves, Bull & Lakin built extensive sidings at Wilmcote Station and a bridge for the quarry tramway to cross over the canal. The overall effect of this gradual move to rail transport was to impoverish the boatmen, and they had to relinquish their houses and move onto their boats.

Other local companies continued to use the canal. Kendal's, the company that made brewing syrup for the local Flowers Brewery, continued to have their supplies of glucose chips delivered from Bloxwich by boat. Every boat was followed by a crowd of children begging for lumps of the 'Chemical', the only sweet treat most of them had ever tasted. The company that made the barrels for Flowers, Cox's, also continued to have supplies of timber brought by water.

The arrival of the railway had caused a geographical shift in the location of the town's manufacturing areas. Cox's timber yard had moved from the One Elm down to the canal basin as its business thrived. There it was furnished with canal and river wharves as well as sidings from the tramway. Many other businesses followed suit to create a massive industrial and transhipment area that was to prove to be too far from the railway. Once the Stratford Railway opened its first station close to the One Elm, most started to drift back northwards. Isambard Kingdom Brunel had looked over the tramway for the Oxford, Worcester & Wolverhampton Railway after they had bought it and suggested that the only section suitable for conversion to normal steam-powered operation was the section between Moreton-in-the-Marsh and Shipston on Stour. Flowers Brewery and Kendal's remained beside the canal close to the One Elm. The town's brick works moved from the south of the canal to the north, partly to exploit new seams of red clay and partly to be on a convenient level for a rail spur across Birmingham Road directly into the works. The old clay pits became a useful borough rubbish tip.

Waterside, with its warrens of warehouses and sidings, started a period of decline as the traffic moved away from the canal. The two pubs, the Anchor and the Ship, catered for the bargees and wharfingers. The Ship gradually got a fearsome reputation as a house of seriously ill repute.

Without regular trade Stratford's once thriving port became derelict.

Kendal's continued to received their loads of glucose chips from Bloxwich by boat. This picture from 1910 shows what appears to be a Severn Canal Carrying Co. boat unloading beside the company's original factory at Lock 52. Once again the quality of the canal maintenance under GWR ownership seems adequate with the lock gate and paddle gear in good condition.

Boats passed through the lock onto the river only rarely.

Possibly the very last canal boat to use the River Avon was photographed in 1882. Lucy's Mill below Holy Trinity Church transferred virtually all its cargoes to rail when the East & West Junction Railway opened. No doubt a few cargoes were still available for boats until Lucy's Mill lock finally failed. The boat Avon was owned by Mr M.C. Ashwin of Union Street, Stratford, and ended up sinking in One Elm Basin in 1909. (Warwick CRO, ph195/7)

The reputation was such that when Emma Bradley tried to sue Charles Beckett for the paternity of her child, the case was thrown out when it was asserted that she had been seen in the Ship. The Warwick Tavern gradually lost its customers as the bargees became poorer, and John Tallis closed its doors and sold up in 1855. Richard Timms, a coal merchant bought the property, and presumably brought in his coal by boat, and sold it to the townsfolk with a horse and cart.

The poverty was by no means confined solely to the boatmen, many working class citizens of the town were living in abject squalor, barely able to support their families. The canal found another and far sadder use, in 1858 no less than three new-born children were found drowned in it.

Stratford Gas Works began receiving its coal by rail almost as soon as the tracks were open. Their shift was not so dramatic since the first station was designed, in part, to supply them a mere hundred yards up the canal. The regular, if low volume, traffic on the canal ensured that the GWR had a reason to ensure the maintenance of the waterway. This was a permanent problem because the original construction of the canal had been done on such a low budget that much of the structure was crumbling away. The original brickwork of many locks and bridges was suffering frost damage. Edstone Aqueduct required a lot of work – the brick piers were exposed to both wind and frost, as well as the early morning sun, and the bricks on the east side were eroding away to the point where the weight of the water and iron trough might collapse. Fortunately the bricks the GWR used to repair it were of engineering grade and have stood the test of time. Almost every lock shows evidence of substantial repairs with this type of brick.

During the 1870s, with the battle of the gauges well and truly over, the GWR used their old broad gauge rails to make mile posts and fence posts. The original ones, being made of wood, had rotted away. The new mile posts were set every $\frac{1}{4}$ mile to give the toll collectors an exact distance on which to levy the tonnage rate.

One additional source of income for the railway company was to rent out the lock-keepers cottages. Whilst the canal company owned them, they had been solely for the use of canal staff, and left empty if there was a staff vacancy, but under railway ownership the extra rents helped defray the expense of keeping the navigation open. As the number of staff was reduced, so the cottages became empty. Only a few of the cottages, at strategic intervals, were kept for canal staff, the rest were rented out. As housing stock they hardly commanded the highest price – the barrel roofed ones were diabolical to live in since the roof leaked, and the rest were simply one up one down hovels; still they were adequate for many of the farm labourers in the district. Quite a few gained unauthorised extensions made from the bricks cut out of the locks when they were repaired, or occasionally, GWR engineering brick. The tenants were still better off than the boatmen.

As the boatmen's conditions worsened, some more publicly spirited souls attempted to introduce legislation defining their working conditions. Various Canal Boats Acts came into force during the 1870s to try to ensure vaguely decent standards of living. Their conditions were truly awful, with whole families crammed into a tiny cabin. Literacy was out of the question for children who never stayed in the same town for more than a day or two, and the overcrowded cabins were inadequately ventilated, leading to very poor health. In 1879, the 1877 Canal Boats Act came into force. It required the local authority to provide school places for the children on boats registered in that town, and that boats be inspected as to their suitability for the families living on board and their general condition.

When the Act came into force, sixteen boats were considered resident in the town: three belonging to Hutchings & Co., local general carriers; four to Greaves, Bull & Lakin, the Wilmcote Quarries; two to Mr Rainbow of the Anchor Inn, Waterside; one each to Messers Ashwin, Court, Dixon and Bullivant; and one to the boat builder Thomas Farr of the Warwick Road boatyard, which was to be lent to clients whilst he repaired their boats.

The river was rapidly becoming un-navigable.

The river at Marlcliff, near Bidford.

Mr Farr took most of his trade from repairs rather than new boats; Greaves, Bull & Lakin ordered six more new boats from him over the next decade, but mostly he was busy repairing and painting the narrow boats so that they complied with the Act. The dry dock and yard have long since vanished beneath new housing, although his house, Bridge Cottage, remains.

As the century wore on, the shift in population became more and more marked. The town streets where all the boatmen lived, Shakespeare and Mulberry Streets, housed industrial workers instead. The canal basins changed again. The tramway tracks that wove through coal and timber wharves were lifted, leaving a small core of transhipment businesses clustered around the original basin. The GWR, never a company to let anything go to waste, used the old track to make racks in which to store the stop planks for making temporary dams in the canal. The dry docks of the boat builders gradually silted up through lack of use, the two in the second basin were closed, leaving only the one near the Warwick Road. The second basin, in front of the original Memorial Theatre, was tidied up and given a few flower tubs. The GWR sold it to the Borough of Stratford, who promptly demolished the ramshackle collection of warehouses and yards so that the Memorial Theatre would have a far better outlook. The main basin remained for coal and other goods to be moved from the canal to the tramway, but in smaller and smaller amounts.

The river navigation had virtually collapsed by the end of the century. Complaints about the lack of maintenance had started as early as 1863 and by 1873 the only through traffic left was the steamer *Bee* bringing corn up to Lucy's Mill from Tewksbury. In June 1875 the GWR gave up collecting tolls, which enabled them to give up responsibility for the navigation, and very shortly afterwards the navigation was vandalised by persons unknown, bringing virtually all boat movements to a halt. Although the councils of both Evesham and Stratford took the GWR to court to force them to reopen the river, their attempt failed. As the century closed there was a brief flurry of interest on the Lower Avon, but the Upper Avon slid into complete dereliction.

Deprived of its through route capability, the canal traffic declined still further. By the turn of the century both river and tramway were unusable and, although the main structure of the canal had been maintained, the waterway was starting to silt up so that boats could no longer pass along it with a full load. At the dawn of the twentieth century, the Stratford Canal was in a plight typical of all railway-owned canals, starved of funds and freight it could do little more than cling to life, weeds choking the channels and a scene of poverty for all those working the system.

Eventually the lock gates rotted.

The lock became overgrown and the bottom gates finally collapsed in 1885.

Wilmcote flight at the turn of the century. Even then the canal could easily accommodate fully laden boats.

Lapworth Flight, the northern section of the canal continued to have regular traffic.

A typical Stratford Canal split bridge. The gap in the centre allowed the tow rope to pass the bridge without interruption.

The era of rail transport had arrived. This East & West Junction train is leaving the station at Lucy's Mill. The railway itself was never a financial success and was absorbed into the LMS after the war. The railway provided Lucy's Mill, in the background, with an alternative transport route to the river.

50

In Stratford itself the second canal basin was tidied up and landscaped to provide a more aesthetic setting for the theatre.

The original basin continued to serve the ever dwindling number of boats. The few buildings that remained were used in the main by businesses that had no connection with the canal.

Four
Retirement with God's Wonderful Railway

As it began, the twentieth century did not hold much promise for the Stratford Canal. Its owners, the Great Western Railway allowed traffic to use it, but spent the absolute minimum on maintenance. Elsewhere around the country the same scenario was being played out, rural canals simply couldn't pay their way and were gradually heading towards extinction. Industrial concerns like the Birmingham canals were secure, but the Stratford was most certainly not. The only thing that kept them going was not the altruism of their owners but simply that it would cost more to officially abandon them than to leave them alone.

Greaves, Bull & Lakin, the Wilmcote quarries, decided to move their main activities to Long Itchington and Stockton, and the Wilmcote Quarry closed in 1908. At a stroke the canal traffic on the southern section was halved. There were still kilns at One Elm Wharf, but they were cold and empty, the boats tied up beside them were idle. Mr Farr, the boat builder, lost most of his trade and it seems he retired to a nice new house in West Street. Even the coal merchant in the former Warwick Tavern had given up getting his coal by boat. With no resident boat builder, Stratford Borough decided to close its register of boats. By 1909 there hadn't been a new boat registered for nineteen years. The closing summary of the register gives details of the fate of all the boats. One was still trading for Greaves, Bull & Lakin, of the rest, fifteen had been sold off and removed, three broken up, one sunk in the Bancroft Basin, one sunk in One Elm Basin and one described as 'taken to ? [sic] by Boat Builder (who supplied new one) & supposed to have been broken up. This transaction happened 20 years ago therefore this boat is doubtless a thing of the past.'

On the northern section traffic still moved comparatively frequently since boats could use the route as a short cut from the Warwick & Birmingham Canal into the south west of the Birmingham area. However the eventual subsidence of Lapal Tunnel on the Dudley Number Two Canal reduced the number of destinations. The whole network of smaller canals was in terminal decline. A Royal Commission had been appointed in 1900 to investigate the state of the system but although it came up with a variety of ideas, ranging from complete Nationalisation to a fairly drastic pruning of the existing network, nothing tangible came of it; certainly nothing relevant to the existence of the Stratford Canal.

Before the First World War the canal was in fairly good shape considering the backlog of maintenance. The water levels in the long pounds were kept at weir height in all but the worst drought. The children in the cottage at Lock 40 could always tell when a boat was on the way down from Bearley Lock, four miles away, because of the surge of water into the by-weir. This was important knowledge if you were going to be the first to get to the boat and scrounge some of the glucose off the crew. The boats were so infrequent that a lot of waterweed grew, but the banks and hedges were kept in tidy order. Using the Top Gate of Lock 40 as a short cut to school was not such a good idea, young Malcom Green apparently had an epileptic fit crossing it and drowned in the lock in 1907. It was a traumatic event for all concerned, Mrs Pickering, who lived in the cottage, constantly reminded her children of the danger ever afterwards. As with most cottages, the one at Lock 40 housed not a canal employee, but one of the local quarry workers. The little 14ft square one-up/one-down had been extended by 10ft to add a scullery and an extra upstairs room. Life was not that much more luxurious for this, the whole family still slept in one room, four large straw mattresses arranged on the floor around the walls, a single range downstairs in the front room. Despite these atrocious conditions Mrs Pickering still had the compassion to nurse her grandaughter back from the brink. The little girl, Dorothy, had been left in the cabin of a boat all night whilst her stepfather was out drinking, and she got frostbite from the waist down. It took six years of constant care before she could walk.

Boating for pleasure was an inconceivable concept in 1911, and yet one man hired a boat for a tour of some of England's canals. Mr E. Temple Thurston wrote *The Flower Of Gloster* (named after the boat), about a month exploring the Oxford, Grand Junction, Thames & Severn Canals and, luckily for us, the Stratford. The account appears to be written out of fragments of a couple of journeys. A close examination of the text shows descriptions of going up a lock, when his journey should have been taking him down it, but the story is one of the sheer joy of cruising through an English spring in the heart of the country. The Stratford Canal inspired him to write more about beauty than the landscape, or more particularly what the locks and bridges were like, how deep the water was, who the lock-keepers were, and so on. The descriptions of the canal, shortly before Europe was to be devastated by war are priceless, and perhaps he had some premonition of the chaos looming on the horizon. The canal left him with the indelible impression that the route was right out of the track of things, a self-contained world one step removed from reality.

'How can a man care for that devastating march of civilisation when it means the ultimate destruction of such places in the world as this,' he mused to himself as he descended the locks from Kingswood. Surrounded by trees in blossom he sang away to himself all the way to Lowsonford. 'It is an event, unparalleled almost in excitement, when a barge comes through to Stratford. Then all the little boys and girls rush down the street to the old red stone bridge to watch it as it passes through the lock.' He didn't know that half the boats would have been carrying those tasty glucose chips.

Reaching Stratford, Temple Thurston says he donned his knapsack and walked to Tewkesbury to rejoin the *Flower Of Gloster* for his trip along the Thames & Severn. He wasn't over-enamoured of the town, a vision of Marie Corelli in her gondola seems to have erased every other memory.

There is no mention of running aground, or getting stuck in collapsing locks, so presumably the canal was only plagued with the waterweed. That it was deserted is incontrovertible. Stratford, as a town, had to keep a register of boats, their condition and occupants. Some companies, like Fellows, Morton & Clayton, went as far as building an extra cabin in the bows of the boat to comply with the Canals Acts. Most did not, and an inspector checked the boats from time to time. He didn't have much to do in Stratford. Nine boats were inspected in 1911 – four belonging to Fellows, Morton & Clayton, the rest to the Severn Canal Carrying Co. – all, it appears, bringing glucose chips to Kendal's.

Flowers Brewery moved from their canalside premises in Brewery Street to a modern factory on the Birmingham Road which was more convenient to the railway and, increasingly important, the main road. Kendal's moved into the old site and continued to receive their glucose by water. Here Hanley and Sandbach are unloading in 1914. These were solid, well-built boats run by Fellows, Morton & Clayton.

Kendal's moved their factory from the rather tatty warehouse beside One Elm lock in about 1911. The company archive has one photograph of a delivery c.1910, a rather battered and nameless boat, almost certainly belonging to the Severn Canal Carrying Co., being unloaded whilst it was in the lock. It was a practise that would have driven the first lock-keepers wild since it blocked the navigation, but made unloading vastly easier. If there had been any other boats around, there would have been trouble. The traffic was now so low that it just wasn't a problem. The Flowers Brewery had moved up Birmingham Road to a brand new brewery, leaving the old one vacant. Kendal's moved into the old brewery, which had a perfect access to the canal for loading, and carried on making their brewing syrup. In the spring of 1914 another pair of boats was caught by the company photographer, both were built with wrought iron sides and wooden bottoms, complete with fore cabins so that they could house a family, in compliance with the Canals Act. The captain, Mr Price, and his family were lucky to work for such a responsible company. Both boats were built at the turn of the century, *Hanley* is still going, *Sandbach* was terminally involved in the blitz on Birmingham. It was a busy year for the inspector, thirteen boats underwent his critical eye, and all passed with no problems.

The First World War came, and went, leaving the canal untouched. The slaughter of a generation created a massive shortage of labour and the whole canal system took another lurch down the slope to annihilation. Even the town's Canal Boat Inspector gave up checking the boats from 1915 to 1919. A new threat appeared as hundreds of these new-fangled war surplus petrol-powered trucks started a new form of transportation. Road haulage may have been in its infancy but it was to grow up to threaten both canals and railways. The very last boat, *Banbury*, registered in Stratford was sold off in 1916.

The tramway was now completely unusable, the passing loops had rotted away years before, their wooden rails weren't replaced, but with the massive shortage of iron caused by the war, the rails were lifted and melted down in 1918. Only the section from Moreton, north to Shipston on Stour, which had been converted to steam power in 1889, remained in use. The traffic on the northern section dwindled away as the GWR decided that it would be a waste of money to convert it to steam power since the routes into the town were already well served by the branches from Honeybourne and Hatton. The lack of maintenance had meant that it was difficult and awkward to use. The last commercial use of the line had been around 1902. The rails, lightweight as they were, were worth more as scrap. The route out of town was left as a childrens' playground and lovers' lane.

Traffic down the canal dwindled away. Inspections rose through the 1920s, though this is no indication of the total traffic, and peaked at twenty-eight inspections in 1932. The inspector seems to have been on the ball right enough, and all the boats were Fellows, Morton & Clayton craft. These boats were big, and needed a waterway in fairly good condition to be able to get through. However the canal was steadily getting worse. The coal merchant, Mr Platter, at Wilmcote used to take his heavy horse up the towpath to help drag the glucose boats over a particularly shallow bit. Payment in glucose chips was ideal for making home-made wines. It's nothing short of a miracle any of the stuff ever got to Kendal's. Mr Platter's son, Jim, recalled the bargees, an old man and his two massive daughters, who had voices like foghorns, could out drink anyone in the Masons Arms and swore like troopers. Their names are lost in the mists of time, could they have been the two demure little girls sitting on the boat *Sandbach*?

The inter-war years saw the total decline of the navigation. Kendal's stopped getting their supplies by water and the gas works cancelled their boat. The pounds were silting up very badly all along the canal. In 1933 only six boats were inspected, just three in 1934, and finally *Keswick*, in 1935, was the last in the book. The southern section was effectively abandoned by the GWR.

The northern section was badly affected as new diesel-powered narrow boats hammered along them, their wash eroding the banks away. The southern section had only ever had horse-drawn boats along it, but even that was getting too shallow. The lock-keepers dropped stop planks into sections prone to leakage, bodged up the gates with ashes to make it look all right, and mostly toddled off to their allotment. The canal did still need to be kept watertight since the railway was using the water for its steam trains, fed directly from a valve cut into the bottom of Edstone Aqueduct. It was supposed to have had a screen to keep the rubbish out of the engines' water tanks, but a couple of trains were brought to an untimely halt by a fish getting into the works. Stratford Gas Works also required water from the canal and this had to be sent all the way down the line from Lapworth.

In the town centre things were starting to change dramatically. The Shakespeare Memorial Theatre caught fire on 6 March 1926. The fire left it a blackened shell and it was decided that a state of the art theatre would be built to replace it. The new theatre would be bigger and better, and that meant that the site would need to be bigger too. The triangular canal basin was filled in and what wasn't used under the new building was grassed over. The existing hodge-podge of wharf buildings around the first basin was demolished and the whole area landscaped. The site had become so derelict that there were complaints about the smell from the rotting reeds growing up to 10ft high. The last remnants of the tramway were lifted and the GWR left 15ft of track as an exhibit beside the blind arch of Bridgefoot Bridge. They made a cast iron plaque to commemorate the tramway's history. After a century of being the worst bit of the whole town, the waterside started to be gentrified. The GWR sold virtually all the rest of the wharves to the town council in two separate sales in 1932. This comprised a freehold sale of the land around the basin with the exception of an access lane, wharf area beside the Lily Arm and a strip around the edge of the basin and lock about 10ft wide. This area was leased to the council with all the rights of navigation kept by the GWR. The freehold area was sold with a covenant stating that the area must be kept for ornamental use with the exception of the area

Ninety years later, Hanley is still cruising the canals. It has had a variety of jobs and transformations in the intervening years.

to be used for the new theatre and part of the road widening at Bridgefoot. As the triangular basin and half of the Lily Arm were filled in, the level of water in the remaining basin rose to the top of the barge lock gates because the overflow weir was located at the far end of the two basins.

The lock gates, in poor shape when Temple Thurston used them, were getting worse all the time. When a boat did use the canal the temporary seals were broken and the lock-keepers had no end of trouble resealing them and restoring the water levels.

Just before the Second World War one last boat tried to make it to Stratford, and failed. The canal was derelict. The prettified basin beside the Shakespeare Theatre was a slight clue to the existence of the canal, but through the town its route was behind houses out of sight. One visitor to the theatre decided to explore a bit further than the basin, finding his way to the back of the bus station and on to the towpath. He walked north past crumbling overgrown locks leaving the new high-tech theatre behind and immersing himself in a world of dereliction, of stern notices from the GWR and of cottages slumping into piles of rubble. It struck him as an unreal, fantastic contrast. One step removed from reality. Once again the canal had woven its own magic, this curious theatre-goer was Robert Aickman, and his impression of the forlorn Stratford Canal was so vivid and powerful that he resolved to do something about it if the opportunity arose.

Outside the town, the canal maintained its very rural character. Here at Wilmcote, looking north is the bridge on Featherbed Lane, whose imminent collapse triggered the entire restoration movement. (Warwick CRO, ph350/145)

Stratford's tourism industry thrived with the new theatre, attracting visitors to the town's hotels.

Lock Cottage 5 at Wilmcote was home to Mr and Mrs Pickering, here photographed in 1920. Despite the primitive conditions of the cottage they raised a dozen healthy and happy children.

Opposite: *Such bucolic scenes captured E. Temple Thurston's imagination when he wrote of his travels on the Stratford in 1911. His book* The Flower of Gloster *has inspired generations of canal enthusiasts for its accurate description of a way of life that vanished after the first World War.*

In Stratford the river was starting to be used for leisure. The famous author Marie Corelli had a gondola built for the regatta, much to Temple Thurston's surprise. Marie Corelli was renowned for being camera shy, and this is one of the few surviving pictures of the romantic novelist.

Cox's timber yard thrived despite being unable to receive goods from the river wharf in 1910. There are several items of interest in this picture. The crane for unloading river barges is still in place, twenty years after the river navigation had ended. The picture has been taken from the tramway Bridge, looking across the site of the Roman ford to the Georgian Toll House on fifteenth-century Clopton Bridge.

The tramway had become virtually derelict by 1900. The long embankment south of the Avon now forms a pleasant footpath into the countryside and then disappears into an overgrown cutting.

The northern section of the canal remained in constant use. The northern section of the canal has a very different architecture to the south. Originally conceived as a wide waterway with big bridges, when the company came to build these locks, they could only afford to construct them for a narrow boat width. The cost cutting continued on the southern section with even cheaper designs of lock gates and brickwork. The northern section remained in use whilst the southern declined and went derelict. (Warwick CRO, ph352/108/36)

The tramway became a local playground.

An air of rural tranquility descended on the once busy waterway.

In Stratford, Flowers Brewery invested in one of these new-fangled lorries, a direct competitor to both canals and railways. Their first lorries were steam-powered, but petrol ones became available in quantity after the First World War.

The canal's terminus was now disused, and weed grew thickly across it. The reeds grew up to 10ft high each summer and their rotting remains caused many complaints.

The Memorial Theatre burnt down on 6 March 1926; the new one required the gentrification of the Bancroft as well as some of the land occupied by the second canal basin.

The time had come for the demolition of the decayed warehouses and docks. Close examination of the photograph of the work reveals the sunken remains of a boat beside the old canal company warehouse, presumably the one mentioned in the Stratford register of boats.

Road transport was here to stay, and Hutchings coal merchants was demolished to make way for the bus station. Not surprisingly the coal merchants moved to a rail head at the LMS station at New Street. Elsewhere on the canal businesses were looking to the future mode of transport.

Even Kendal's invested in road transport.

Stratford Borough Council tidied up the wreck of the lock at Lucy's Mill when it became clear that the Avon would never be reopened by the Great Western Railway.

The canal didn't appear too awful in places, particularly near the lock-keepers' cottages. Here, viewed from Maidenhead Road, the cottage of Mr Hancox, virtually enveloped in a mantle of ivy, typified the old-world charm of the waterway.

Five

War and Decay

The canal system was pressed into service in the war against fascism along with every other national resource. The northern section of the Stratford saw an increase in use, although the southern was too far gone to be of any use. By 1945 it was impassable and the rest of the rural canal system was in a similar state. The war effort had required Herculean efforts from the railways, which had resulted in a cessation of maintenance on all canals that weren't directly contributing to the war effort.

At the start of the war the waterways network fell into two broad categories, those owned by the railways and those that weren't. The Lower Avon Navigation was in the hands of a group of Trustees and partially maintained. The Southern Stratford was rapidly meeting the same fate as the Upper Avon had when it was owned by the GWR. The railway-owned canals included many that were still functional and these were utilised to the full. A study into the canal system by the chief of London Transport recommended that most of the canal system be closed down, and only those canals that could be of service be transferred to government ownership. The future of the canal system was looking bleaker than ever.

In 1939 an author, L.T.C. Rolt, took his boat on a voyage of exploration around the canal system, such as it was. He felt that this might be the last opportunity to do so. His journey up the Northern Stratford met with all sorts of obstacles from thick weed to lift bridges that wouldn't lift. His account was eventually published in 1944, and struck a chord with many people who felt an affinity with the canal system.

In 1946 Rolt and a handful of other dedicated canal enthusiasts, including Robert Aickman, formed the Inland Waterways Association. The canals and rivers finally had an advocate. The Association was divided into regional groups and they all started to campaign for the retention and improvement of the canals and rivers. Aickman specifically included the Stratford Canal as one that was in urgent need of assistance.

The Lower Avon had passed from several trustees to just one owner, Mr John Whitehouse. Although dribs and drabs of maintenance had occurred, by the end of the war the river had become un-navigable above Pershore. Some pleasure boating was taking place, but the tolls this raised weren't enough to keep pace with the deterioration of the structure. When Mr C.D. Barwell managed to get his boat stuck in Nafford Lock he decided the time had definitely come to do something.

In 1950 the Lower Avon Navigation Trust was formed and bought the navigation. If the government weren't going to do something, and they most certainly weren't, then ordinary people would just have to do it themselves. The Lower Avon started to be restored – money and resources, labour from the Royal Engineers and volunteers from the Midlands branch of the IWA all set to work and gradually the river came back to life.

For the railway-owned Stratford Canal, things were somewhat more complicated. The nations transport system was nationalised with the Transport Act of 1947. Railways and their canals found themselves fighting against the road transport lobby in a battle for survival. Road transport was becoming an ever more dominant force to be reckoned with. The railway-owned canals barely got a mention in the ensuing debates.

The nationalised industry was subdivided into Roads, Railways and the Docks & Inland Waterways Executive. This rather unfortunate grouping of docks and inland waterways did no good at all for the canals. Disputes started immediately about whether or not one should have dockers unloading narrow boats. The general opinion was that they weren't worth the bother and all inland waterway cargo carrying of less than sixty-ton loads (a pair of narrow boats) was simply not cost effective.

The British Transport Commission surveyed the canal network in 1955 to see why the canal system was losing so much money, and what could be done to improve matters. The survey divided the system into three categories: the best, which were still trading actively such as the Grand Union; ones with potential for improvement, such as the northern section of the Stratford; and the remainder, like the southern section of the Stratford. This remainder class were economic basket-cases, either semi-derelict or so little used that they cost far more to maintain than they could ever raise in tolls. This last class contained over 1,000 miles of waterways. The survey proposed selling off or transferring these canals to local authorities, water boards and such like.

If the Docks & Inland Waterways Executive hoped to release themselves from a morass of statutory duties to maintain old canals, they were sadly disappointed. Almost no local authority

The canal at Preston Bagot was weedy, but navigable for a rowing boat. In some parts the lock-keepers had been able to keep the towpath hedges under control.

At Wootton Wawen the cutting was far more difficult to maintain and nature took over with a vengance.

would touch a canal with the proverbial barge pole. They were a liability. Virtually every remainder class canal had become an overgrown mess, used mostly for dumping unwanted rubbish. The canals had turned their back on society at the start of the century, and in so doing gained a very bad reputation. The fairly regular drowning of inquisitive children in urban canals had convinced most people that the waterways would be far better filled in and forgotten. The local authorities knew just how much this would cost, and they weren't going to find the money from their budgets if at all possible.

Warwickshire County Council were offered the option of taking over the southern section of the Stratford Canal, but declined promptly. The owner of a waterway would be responsible for the cost of maintaining not only the bridges, but also the liability for any damage caused by flooding. The cost of closing a canal was high since a waterway has a land drainage function which requires a drainage pipe to be laid in the canal bed before it is filled in. Removal of large and potentially dangerous structures like lock chambers and aqueducts is also expensive. The County Council would have liked to be able to take control of the road bridges, but the cost of such a venture was beyond their means.

There was some activity on the northern part of the canal near Earlswood Lakes where there was the Stratford Canal Club. The boats were quite an assortment of old lifeboats, home-made dinghies and small cabin cruisers, but they all used the canal, and no one there wished to see the canal closed. The IWA had started to take an interest in the canal as early as 1953, publishing a survey of the entire canal, but for the most part the canal had receded entirely from the public arena.

Michael Fox was boating on the Norfolk Broads in 1955 when he noticed a hand-written advert for the IWA in a post office window. Curious, he wrote to the address shown and, as is often the case, was enrolled and rapidly found something to do – in this case, forming a protection committee for the southern section of the Stratford Canal. The remit was to restore the canal to full navigable order for its whole length. A rather ambitious undertaking for half a dozen people, but as David Hutchings, another IWA member, succinctly phrased it, they didn't know that it couldn't be done. Michael Fox initiated a publicity campaign to make the plan known. It was to catch the public imagination far more than anyone could have foreseen.

The Wharf Inn at Hockley Heath was the venue for a meeting of both the canal's protection committee and the Stratford Canal Club on 17 November 1956. The two bodies decided that

Overgrown both above and below the water the canal was not exactly suitable for a narrow boat.

they would work more effectively in unison and promptly combined to found the Stratford-upon-Avon Canal Society.

The Canal Society may only have had seven members, but they were highly motivated. Bearing in mind the 'use it or lose it' proverb, they borrowed a Canadian-style canoe and proceeded to drag and paddle it up the canal from Stratford to Earlswood. It wasn't a journey for the faint hearted, fortunately the weather of February 1957 wasn't too cold, but there were plenty of other obstacles. Some parts of the canal were just solid reeds, Wilmcote flight of locks was mostly dry, the Great Western Railway had bypassed the whole flight with a pipeline to convey water down to Stratford Station, and allowed the locks to disintegrate. The lock cottage at Lock 54 might have been fine, but the one at Lock 51 was just a pile of bricks, at Lock 50 the cottage was rapidly collapsing as was the one at Lock 40. The canoe had to be dragged up the entire flight along the towpath. Once beyond the parish of Old Stratford the channel improved and for long stretches it was possible to paddle along without too much hindrance.

The journey took six weekends, but it was an official use of the canal. Michael Fox had bought the correct toll ticket, at 2p per mile. It was to prove to be a very wise investment.

With this hands-on experience, the Canal Society approached the British Transport Commission and suggested a programme of voluntary work on the canal to improve it enough to have more regular use for canoes and other lightweight boats. The BTC replied in a rather vague way, suggesting that everything should be put on hold until the current government inquiry was published. They would consider a report on the canal by the society, and would discuss the canal with them, but that was all. The report was duly sent in, highlighting the risks a leaky derelict canal posed, and proposing that volunteers from the society could help patching up the holes and clearing the waterweed where it blocked the overflows. The report vanished into the labyrinthine depths of British Transport on 9 May 1957. The local engineer of the BTC was refusing to allow larger boats access to the canal, as Christopher Clifford, an IWA member, discovered. Legally, the BTC should have allowed any boat to navigate the canal, and they were legally obliged to maintain it in a navigable condition. These niceties had gone by the board and the truth of the matter was that the BTC wanted rid of most of the narrow canal system.

The Canal Society was starting to attract more members as Michael Fox's publicity campaign got into gear. The offers of volunteer help were all covered by the local press, as was the first general meeting in March 1957 and a small boat rally at Hockley Heath. The membership gradually increased to nearly thirty. They had a lot of work to do, some of which was preparing evidence for the government inquiry taking place, the Bowes Committee. Their principal aim was to convince the committee that although the canal may no longer have a role in the industrial freight transport world, it could have a strong and healthy future as a leisure resource. Perhaps more importantly, the canal society's report suggested that if the canal's future could not be assured by a national body, perhaps a local organisation might be allowed to take over the concern. The IWA was also promoting new use of old canals along with the Kennet & Avon Canal Association. The time to consider canals not as freight transport but as a national asset had come, if only the government could be convinced of this. The Stratford Canal Society's contribution to the debate went in late, but at least it was allowed and the Bowes Committee retired to consider the inland waterways' future on 11 July 1957.

In August the reply finally came from British Transport, flatly refusing to allow any volunteer work until the result of the inquiry was released. Something of a set back, although not entirely unexpected. The principal motivation behind any large nationalised bureaucracy is not to get anything done, but to look busy behind a desk. Any idea even slightly unusual was viewed with deepest suspicion.

The canal society and the IWA carried on with the press campaign, gradually turning public opinion from a harsh opposition to at least a vague tolerance. Stratford Borough Council was certainly not convinced however, and maintained that the whole canal would be better off filled in and sold off, the only problem being the cost. Finally in February 1958, long before the Bowes Committee had come to any conclusions, the decisive moment arrived; Warwickshire County Council decided that it would apply to Parliament itself to have the canal officially abandoned.

Wilmcote flight was in very poor condition. Here, looking down from Lock 48 it doesn't appear to have been too badly vandalised, but by now the cottage at the bottom lock had been reduced to a shell by local kids. The cottage at Lock 51 was nothing more than a pile of bricks.

The canal had been founded with an Act of Parliament, and it needed another to be wound up. The County Council needed to upgrade several roads to take greater weights, and the canal authority was only bound to keep their bridges to the weight loading originally specified. For most road bridges this was five tons, which was routinely exceeded with the result that many of them were starting to collapse. Featherbed Lane Bridge in Wilmcote had got to the stage where it was only held up by huge baulks of timber. The County Council would have to pay to build a bridge capable of taking the greater weight of lorries, or if they could get the canal abandoned then they could save thousands of pounds by culverting the canal and not bothering with a bridge at all.

The canal society was suddenly galvanised into an all out battle for the survival of the canal. Letters of protest went out in every direction and a public meeting was arranged at Stratford Town Hall for 26 April. Help was sought from the IWA and other sources, including the National Trust. The Trust was already committed to preserving many historic houses and structures and was considering preserving some historic canals because the prevailing political climate seemed to indicate that they might all get filled in and lost. The Staffordshire & Worcester Canal was one candidate, and so was the southern Stratford. The thought was that if one of these could be saved, this aspect of the Industrial Revolution would not be entirely lost. On 11 April the Trust approved the principle of preserving a few selected canals, but only if they could be self financing. The National Trust was in a difficult position; John Smith, the honorary assistant treasurer of the Trust, came up to Stratford to meet Robert Aickman, Christopher Clifford and David Hutchings. They discussed the various options available and

Even a cursory inspection showed that the gates were rotten. The brickwork in the chambers appeared to be in good condition in 1959.

Timothy's Bridge, on the outskirts of Stratford, was collapsing. Huge cracks had appeared and the fields around had been earmarked for the development of an industrial estate. When the bridge was finally replaced with a concrete road bridge, the cast iron split bridge managed to vanish. If anyone knows where it went, the local canal society would like to know.

walked along the canal. The Trust was in the position to give the canal restoration credibility, but couldn't be seen to be part of a protest movement.

The protest meeting at the Town Hall was a resounding success, with more people turning up than there was room to seat. This may have been helped by the fact that that morning Stratford had woken to find a small boat floating in the middle of the basin with a banner proclaiming 'Save The Stratford Canal' for all to see. The meeting saw the start of the active phase of the restoration of the canal, with £1,000 promised for a restoration fund. Hopes were raised that a full restoration was not only possible, but imminent.

The County Council carried on with its application for abandonment, sending the paperwork to the Minister of Transport on 12 June. Their excuse for this was that there had been no boats using the canal for three years, and thus it was disused. They were unaware of a certain canoe trip.

A protest rally of boats was organised by the Coventry Canal Society on 5 July; they were all small boats and vital in proving that the canal was partially navigable and well worth saving. The exercise was designed to achieve the maximum publicity, and again succeeded magnificently. The obstacles were mounting up against the canal though; on 15 August notices were posted all along the waterway proposing that the canal would be closed on 15 September. If anyone wanted to object they needed to contact the Ministry of Transport.

There were a lot of objections. The National Trust objected that the proposed closure was premature as they were still in negotiation with the BTC about a transfer of ownership and restoration. The canal societies printed out thousands of objection forms and thousands of people used these to register their opinion. One in particular came from Stratford Canal Society; how, they argued, could the County Council claim that the canal was disused for the

The cottage halfway up Wilmcote flight remained in good order. Since the lock-keeper still lived in it there was no vandalism from the local children, although they used to 'borrow' his punt from time to time.

last three years when they had officially used it with a canoe, and had the toll ticket to prove it. Shortly after this the ministry asked for the toll ticket to check this very material fact. That one little bit of paper was probably worth more than the opinion of over 6,000 people.

As 1958 drew to a close the National Trust started serious and detailed talks with British Transport about the future of the canal. The prevailing opinion of the transport authority was that the waterway was a liability, and if left to them their cheapest solution would be to fill it in. If the National Trust wanted to take it over, they wouldn't object. In fact they would prefer this as then they wouldn't be seen to be destroying something that they were supposed to be looking after. These talks were held during the period that the Ministry of Transport was considering the canal's closure. Certainly the National Trust knew that the County Council's application for abandonment was almost certain to fail because of the toll ticket, and proceeded to arrange to lease of the canal at a nominal charge from British Transport.

The National Trust needed to work with the co-operation of all the statutory bodies relevant to the canal if the canal's future was to be secure. The County Council, although having started the process to abandon the canal, were persuaded to pay for new bridges if the canal was going to be used again. Stratford Borough Council were far more suspicious of the whole project. They had spent years enduring complaints about the state of the canal from residents, and there was still a deep seated fear of itinerant bargees appearing and wreaking havoc. They really didn't want a convoy of grubby old coal boats moored in front of the theatre frightening the tourists. The possibility that gloriously colourful pleasure craft might one day enhance the very centre of the Shakespeare industry hadn't occurred to them and, to be fair, such boats wouldn't be built for another decade. Converted coal boats and wooden cabin cruisers were the main stay of pleasure cruising. Stratford Council were adamant in their opposition but they only had a certain amount of jurisdiction over the canal, particularly around the basin in front of the theatre.

Lock 49 in a bit of a state, to put it mildly. As the gates failed and the pounds dried out, luxuriant plant growth started in the fertile silt.

Lock 48. At first the restoration team thought that they would be able to use quite a few of the original lock gates.

Lock 46. The lack of maintenance during and after the Second World War had effectively closed the canal.

The canal society and other volunteers had to wait, both for the National Trust's negotiations and the abandonment process to run their course. They couldn't even get on with any work on the canal's structure, and especially not start any more protests that might further incense the local authorities. However as the talks proceeded towards a conclusion, permission was given for a limited amount of towpath clearance during the last month of 1958.

The struggle to keep the canal alive, if not actually navigable, was drawing towards a resolution. In March 1959 the canal society and other volunteers were allowed to extend their range of activities to include clearance work in Stratford itself. Drawing on the experience of the Lower Avon Navigation Trust, they managed to obtain help from the Territorial Army and set to work on the pound above Lock 54, beside Mr Hancox's lock cottage below Maidenhead Road. The TA brought a drag-line dredger with them and an all-out blitz on this eighty-yard length worked wonders. A highly visible stretch of canal was turned from a foul-smelling rubbish dump into a tidy navigable waterway, even some of the town's councillors were impressed. It was a practical demonstration of the merits of reopening the canal and the implications were felt further afield than the local councillors. The new Inland Waterways Redevelopment Committee, who were charged with planning the future of the 'remainder' class canals, came to inspect the work and decided that volunteers could indeed play a significant role in the future of the canal system.

The pound below Lock 54 was the next to receive some attention. The lock itself wasn't in too bad a state; it had been spared vandalism because it was right beside an inhabited cottage. The result was to make the whole project appear a lot easier to achieve than most sceptics supposed. The volunteers moved on to the pound below Wilmcote flight, trying various means to clear the thick weed that had choked the channel.

At long last, on 22 May 1959, the Ministry of Transport announced that the County Council had failed to prove that the navigation had been disused for the statutory three years and so the application for abandonment was invalid. The navigation was saved.

What was to become of it now?

The rest of the flight was in a similar condition. It soon became clear that very few of the original gates could be used; most were completely rotten.

The lock-keepers prevented vandalism to the canal near their cottages. This is Cottage No.4, halfway up Wilmcote flight. Quite why the GWR numbered the cottages from Stratford when the bridges and locks are numbered from Kings Norton remains a mystery.

Lowsonford Bridge. The arch of the bridge is designed for working horse boats, and thus is very low, as many modern narrow boaters have discovered to their cost.

The cottage at Lock 40 was nearly derelict. Like virtually all the lock cottages it had no amenities and very poor access, luckily this cottage was so far away from the town that it remained un-vandalised.

The locks in the flight closer to the town were vandalised.

The cottage at Lock 54 still housed the lock-keeper, and the canal nearby seems in tolerable condition. The lock-keeper was close enough to ensure that the gates could be kept watertight with ashes and turf even though the wood was rapidly rotting. No doubt the silt in the empty pounds would have smelt awful in the summer, so keeping the water levels up made his life a bit more pleasant. The stop plank rack beside the hedge on the left of the picture is one made from the old tramway rails by the Great Western Railway. The young lad standing on the lock balance beam is none other than the author, expressing an early interest in waterways.

In the countryside by Preston Bagot the canal had quietly slipped into history. Without a lot of work the canal could easily have vanished without trace.

The decaying locks were being reclaimed by nature. This is Preston Bagot Top Lock.

The canal bed slowly becoming no more than a damp ditch.

The work before the restoration at Lock 54. The original design of the canal was to have the by-weirs in brick-lined tunnels to carry surplus water around each lock. These culverts were virtually all blocked when the canal started to be restored and nearly all of them have now been opened out to make maintenance easier.

Even this apparently good pound needed dredging. The accumulation of silt can clearly be seen as the dredger works upstream. Dredging uphill like this allows the water to drain away from the work site, unfortunately for most of the canal the dredgers had to work downhill, within a pool of water and thus unable to see how well they were working.

The early restoration was carried out in primitive conditions. Today's restorations have far more Health and Safety legislation to take into account.

The restoration team and British Transport staff had to explore the state of the locks to assess what would be required.

The cottage at Lock 40 was lucky to escape demolition and was partly restored by Everard Berry, one of the volunteers.

For the first time in decades, the stop planks were put into the top of Lock 28. In many locks ground movement meant that they no longer fitted.

Above: Once a pound was drained, just how dreadfully silted it was could be revealed.

Left: A surprise wreck emerges from beneath the dappled water, lost and forgotten for fifty years, this boat won't be carrying any more cargoes.

Inside the locks of Wilmcote flight the damage to the walls was more severe than originally anticipated. As the locks had been empty for so many years, the pressure of the ground each side made the sides gradually collapse inwards. At first the damage didn't appear to be that severe, but the cold winter of 1962 made the situation worse.

Six

Restoration

What was it like? My parents warned me never to play near the canal, so I got to know it pretty well. The house in which I was brought up had 'Good Stabling' written down the canal side wall, a glass door etched with Snug on the dining room, a cellar full of ancient barrels and bottles, even a dozen stables outside. It had been a bargees pub; though now closed since there weren't any bargees left. Once it had been the Warwick Tavern, built by John Tallis in 1831, once a coal merchants and once the very first Catholic school in Stratford. It was a bit battered by 1961. Every time it rained heavily the canal made its way into the cellars via drains that should have drained into the cut rather than vice versa, an unwelcome result of the basin's overflow being blocked in 1932.

Next door to our rambling old wreck of a pub was the boat builders house, Bridge Cottage. Its garden then contained the remnants of a dry dock, silted up to ground level with only a few coping stones left to show the whereabouts of Mr Farrs' workplace. The silt did grow absolutely phenomenal rhubarb, taller than I was at the time. The pound between the Warwick Road and the basin was heavily weed infested, but with enough clear water to allow Oscar, the king swan, to have his nest there. Oscar was a fearsome bird quite capable of doing a six-year-old some serious damage, so that bit of the towpath remained thoroughly out of bounds. North of Lock 55 was Mr Hancox's cottage, a neatly kept little house where the remaining canal employee in the town lived. He didn't seem to have a great deal to do and didn't mind stopping for a while to chat with curious little boys. The lock at Great William Street had had its side cut away to allow flood water to by-pass the existing blocked by-weir, by running down the towpath under the bridge. Not to be trusted if there had been a lot of rain, there was a fair torrent flying down there. The canal above here was virtually choked by reeds, beyond the railway bridges and a set of air raid shelters were fields and a cranky little iron bridge up to Bishopton. It was very difficult to see water, just extremely squishy reeds. The locks of Wilmcote flight were completely dilapidated with their gates smashed, leaning at crazy angles across the brick chambers. In one of the side pounds the cabin of a boat protruded over the vegetation, the rest sunk bow down into the mire, dark water lurking across the floor and bed inside the cabin.

There was a tangible air of other-worldliness about the whole canal, the same feeling that had so struck Temple Thurston and Robert Aickman, struck me. Can a canal have a 'Genius Loci'? I believe so. There was this air of a quiet anticipation, as though the ghost of every boatman was watching in sorrow at the canal's dereliction. It always seemed quieter on the towpath than it should have been, the town's noisy traffic slightly muted.

The negotiations between the National Trust, Inland Waterways Association and the British Transport Commission went ahead well. The IWA and the local canal society had to keep a low profile though. The IWA, in particular, had protested extremely vociferously for the retention of the entire canal system and were regarded by many civil servants as a nuisance. Never the less in the Transport Bill of 1960 the proposal to lease the southern section of the canal to the Trust was included, with an option for the Trust to purchase the freehold if it desired. The Bill was passed and on 29 September 1960 the National Trust found itself the proprietor of the South Stratford Canal.

David Hutchings was appointed Canal Manager. David was an architect by profession, living on a narrow boat. He was heavily involved with the IWA, being chairman of the Midlands branch. He had a fairly brisk way of dealing with official obstruction, a typical example was when Robert Aickman and he were trying to cruise down a flight of locks in the Black Country, the BTC engineers padlocked the gates closed; a hacksaw opened them again. He was going to need every ounce of his resourcefulness to get the Stratford Canal open again.

Restoration or reopening was, and still is, a bone of contention. The National Trust wanted the canal to be restored to something like its original condition, but this was to prove to be too costly. The budget for the work was £42,000, which assumed a fair proportion of the labour would be done by volunteers. The time scale was set at three years. Under these constraints it soon became apparent that putting the canal back to its original condition was out of the question. Preliminary surveys showed that virtually every lock gate would have to be replaced, and quite a few locks on Wilmcote flight would need rebuilding. Right away the design for the new gates departed from the original, using part-seasoned oak, with lighter weight planking and balance beams. Work started on Lock 23 to see whether the new design would be practical even before the official start of the restoration project.

The official start of the project was 1 March 1961. An appeal for funds had raised £20,000 and the Ministry of Transport promised the additional £22,000. David Hutchings moved his boat *Ftatateeta* down to Lapworth and got cracking, working virtually every hour of daylight. He needed to, the next three years had to see the dredging of a quarter of a million cubic metres of mud.

In 1961 restoring a canal was not the kind of job that had ever been tackled before. There were no instruction books, the only useful source of information was the Lower Avon Navigation Trust who were three quarters of the way to completing their project. The Lower Avon had successfully utilised help from a variety of unusual sources, the Territorial Army, the Royal Engineers, volunteer labour and even not so volunteer labour (turn up at a lock in your boat and the chances were that your days cruise would end prematurely and you would end up rebuilding the lock you had hoped to be through in ten minutes). There was a great deal of discussion about tactics between the Lower Avon Trust, the IWA and the National Trust. David Hutchings' vigour and determination utilised this small reservoir of experience and expanded it.

The work was divided into roughly three sectors. The first was the clearance of the jungle that had grown up along the towpath and around, sometimes inside, the locks. Almost all of this bramble bashing was carried out by a hard core of volunteers wielding saws and slash hooks. These work parties travelled ahead of all the other works so that there was room for the dredgers and lock gate fitters. The second part of the work was dredging out the accumulated silt and debris. In the early 1960s hydraulic-powered diggers were unknown and the National Trust had to rely on three drag line dredgers operating from the towpath. Unfortunately the plan of campaign called for the canal to be dredged from north to south, working downhill. Ideally the dredging should have gone uphill, thus allowing the water within the canal to drain away from the point at which the dredging was taking place. The problem was that most of the materials for the reconstruction of the locks had to be loaded onto boats at the canal yard in Lapworth and carried to the work site by water, and this meant that the canal had to be navigable from Lapworth downwards. The dredging was a slow and laborious process. In many places local farmers gave permission for the spoil to be dumped in the fields adjacent, for a small consideration. Where the farmers refused, or the consideration wasn't that small at all, the mud was dumped onto the towpath. It wasn't good for the project since it rendered the path difficult to walk on, or worse, the mud burst its temporary resting place and slumped back into the canal. It also meant that the dredger couldn't drive back the way it had come.

The new lock gates were made at a rate of about one a fortnight at Henley in Arden. At first skilled craftsmen from the British Transport Commission fitted them, but were teaching some of the volunteers the tricks of the trade. The gates were not easy to fit, each one weighing up to a ton, and yet needing to be placed to an accuracy of a few thousandths of an inch. Just getting them to the locks was something of a nightmare, erecting them and ensuring that they were watertight a skill that was nearly extinct. Along with the new gates and paddles, it was anticipated that some of the brickwork in a few locks would need repairing. It soon transpired that far more of the locks would require extensive rebuilding than originally budgeted. This was particularly apparent by the time the work parties started to assess Wilmcote flight. This flight of eleven locks had been built on a hillside sweeping down to the outskirts of Stratford. The underlying ground was partly limestone and partly clay. The locks themselves were built as cheaply as possible in 1815, utilising as many embankments as possible rather that digging into the hard rock below. In the years of dereliction the walls had bulged well away from their original shape.

As the restoration progressed into its second year, it became clear that the project stood no chance of being finished by the date set for its reopening. The Queen Mother had already been booked for the event and the IWA were planning a national boat rally in Stratford. The jungle clearance, dredging and gate replacement were on schedule, but the reconstruction of the lock chambers was way behind. The hopes of being able to rebuild the locks with conventional

Lock 23 was the first to undergo serious repair, even before the hand over was complete.

Mr Clifford of the Canal Society underwrote the cost of a new gate. It was fitted with the assistance of British Transport Commission staff.

brick-faced walls went to the wall, and plain poured concrete onto a reinforced steel frame had to suffice. Other compromises were forced onto the restoration by the tight time scale. Originally the canal had been built with Oxford-style paddle gear. The iron work was very worn, never having been replaced during the working life of the canal. The National Trust obtained new paddle gears from the Wyrley & Essington Canal, which was being closed down. The replacement gear was nowhere near the style of the Stratford, even to needing a different size windlass to operate it.

As the restoration progressed it was gradually becoming clear that it was a restoration of navigation, not a complete physical rebuilding of the waterway and its ancillary structures. Indeed many of the cottages taken over by the Trust were not rebuilt but bulldozed out of the way to make room for the dredgers. The entire character of the canal was changing. The otherworldliness that had so appealed to Temple Thurston and Robert Aickman had been invaded by diesel burning dredgers, the tall reeds concealing and revealing fascinating glimpses of water had gone leaving the canal stark and bare. Time, it was hoped, would heal this temporary mess.

The project received encouragement when the Lower Avon Navigation was reopened in June 1962. The canal's restoration couldn't rely on passing boaters for volunteer help, but work parties from other branches of the IWA came along, and had so much fun with the mud that they became the genesis of the Waterways Recovery Group, a sort of canal flying squad. Some members of the Stratford Canal Society helped on a virtual full time basis, although on the whole David Hutchings was severely short of helpers. The canal society had become very quiet during the restoration process, several members were on the committee set up by the National Trust to oversee the restoration. There were a few issues of the society's newsletter, but for the most part members who were not actively involved in the process were left without any progress reports.

Encouragement was needed in spades. There were times when the whole project seemed threatened. A dispute over access to the canal led to a very tricky moment at Yarningdale. The local farmer refused to allow the dredger to cross his field to get around the aqueduct. The dredger couldn't go back up the towpath since it was covered with a yard-thick layer of mud, and the only way forward was across the old iron trough put up in 1826. To be sure, the original builders hadn't anticipated someone wanting to drive an eight-ton Priestman Cub Dragline dredger across it. They did though.

Problems over access weren't confined to the odd farmer, Stratford Borough Council were still actively opposing the project. The thought of itinerant bargees and dilapidated coal boats mooring in front of the theatre goaded them into trying find ways to stop boats coming anywhere near the basin. First they suggested that the canal should be diverted to a junction with the river north of Clopton Bridge. The National Trust agreed to do this if someone else paid for the work and had it complete for the opening date. So that idea came to nothing. Then there were the battles over access. Under various leases and agreements whereby much of the Bancroft had been transferred from the Great Western to the Borough, there was a right of access from Waterside to a fifteen-foot-wide strip right around the basin and lock. As well as this, the Borough was under a legal obligation to keep the basin navigable; a duty which it had signally failed to do. They had placed a fixed bridge across the lock which prevented boats using it, this had originally been a tramway swing bridge, but the replacement of this was a concrete structure. Then there was the matter of dredging, theoretically the Borough was responsible for this and the National Trust could have demanded the work be carried out. The Borough weren't in the least impressed by all this but the Trust decided that it would not press its claim against them.

Once the new top gate was in place attention could be turned to the damaged brick work. The plan was to replace all the top gates as soon as possible so that the whole southern section of the canal could be re-watered. The brickwork and bottom gates would follow later.

The canal yard at Lapworth was the base for the restoration.

Once the dredgers got to the basin the council had entrenched themselves to the point where they weren't even prepared to allow the mud to be used to help cap over their refuse tip 500 yards away. David Hutchings suggested that left the Trust with only one option; to create a huge pile in the Lily Arm, right at the end of Sheep Street, highly visible from the theatre. The council changed its mind.

As 1964 began there seemed to be an almost insuperable amount of work still to be done. Some of the canal needed to be dredged again, test cruises with a fully laden narrow boat revealed shortcomings in the first attempts. Some locks turned out to be narrower than originally surveyed, as were some bridges. Undaunted the team flung themselves into the remaining obstacles with renewed vigour and by the end of February a cruiser, *Laughing Water II*, managed to get to the Avon. The work redoubled to enable proper narrow boats to make the journey.

By early July an armada of boats was heading towards Lapworth in preparation of the opening and boat rally. The Borough Council decided to use its by-laws to prevent any boat mooring on the riverbanks that it owned, and the National Trust had to agree to prevent boats from mooring in the basin. No doubt there were a few miffed bureaucrats when the theatre offered their stretch of riverbank for the expected 200 boats. The boats started to pour down the canal from 9 July. It was barely controlled chaos, whilst some of the locks had got narrower over the years, quite a few of the older boats had got wider, with the inevitable result. It was probably the greatest usage of the canal since its opening in 1816; water levels dropped alarmingly at the top of the canal, and duly rose even more alarmingly at the bottom.

The Queen Mother was spared the whole journey, embarking on the narrow boat *Linda* at Tyler Street wharf, descending Locks 54 and 55, and entering the basin at tea time on 11 July. Beside her stood Robert Aickman, present for the reopening of the canal that had inspired him to rescue the whole system. The canal was open for business once again.

There was an immense amount of mud to clear from the locks before any work on the structure could start, most of it had to be shovelled by hand.

The sides of the canal were waterproofed with a trench filled with bentonite clay. It wasn't a very successful technique.

Overleaf, top: *Lock 31 required major alterations as well as new gates. It remains a mystery how the original working boats could have got through this lock before the restoration team widened the bottom gate. It was to this lock and bridge that all the children of Lowsonford flocked when Temple Thurston passed through in 1911.*

Overleaf, bottom: *The Royal Engineers provided invaluable assistance. The project manager David Hutchings used every resource he could find to assist the rebuilding of the canal.*

Right: *The lock needed widening so much that it is a mystery how boats had ever got through it. The lock chamber was at a slight angle to the low bridge below it.*

Below: *Lock 30 was much easier in some respects, but as the picture shows, the brickwork of the lock walls had suffered badly and needed extensive replacement.*

Above: A new technique of using reinforced concrete behind a brick facia was used to speed up the work.

Left: This picture of Lock 28 shows the original design of paddle start and gear. Much of this type of gear was retained by the National Trust, but where it was too worn, lock gear from other canals, such as the Wyrley & Essington, was brought in. Today all the old gear has gone.

Drag line dredgers were the main means of clearing the canal. This is a Priestman Cub dragline which was small enough to get into some of the tight spots on the canal.

The 10RB dredger could tackle the larger pounds and work quicker.

Above: *It gradually excavated the silt out of the channel and, wherever possible, the spoil was spread on adjacent fields.*

Left: *The process sometimes had to be repeated several times to get an adequate depth. The fields could cope with the mud, it has a high phosphorous and potassium content, which makes it a good, if smelly, soil conditioner.*

Things didn't always go according to plan. The towpath and embankments had been built long before anything as heavy as a dredger was conceived.

The 19RB type dredger could shift more muck, but needed plenty of space. When the spoil had to be dredged onto the towpath a temporary wall of planks held the mud until it dried enough to be stable.

Dredging work at Lock 38. The canal was originally built to a depth of 5ft and the aim of the dregder was to try to get the canal back to this depth.

Once dredging was complete the canal was once again deep enough to allow boats through. Here at Bishopton Spa, on the outskirts of Stratford, the difference is remarkable.

Right: *The hard structures of the canal such as the locks required far more work than simply dredging out the mud. The left-hand wall of this lock clearly shows a bulge as the ground pressure has pushed it in.*

Below: *The locks had to be pumped out so that work could be carried out on the floor.*

Above: *Large baulks of oak were used to form the lock cill, which the gate closed against to make a watertight seal. The use of power tools was uncommon in the early 1960s so most jobs had to be done by hand.*

Left: *The gates needed to be planed perfectly smooth and flat to ensure a good watertight seal. Modern gates use a rebated rubber seal.*

Right: *It was carpentry on a large scale. The wood was oak which was worked whilst still part seasoned.*

Below: *The locks on Wilmcote flight had bulged badly and required a nearly complete rebuild. The hard winter of 1962 had caused much to the brickwork to bulge in the frost.*

Instead of bricks, the walls were repaired with concrete poured behind shuttering. There wasn't time to repair the brickwork with traditional methods. The huge amounts of broken bricks were incorporated into the concrete to bulk it out, the resulting lumpy concrete was nicknamed Hutchcrete.

Opposite top: *The winter of 1962/63 was one of the hardest since the war. Work on the canal slowed to a virtual stop, as did the rest of the country.*

Opposite bottom: *The canal froze solid, particularly on embankments such as here above Edstone Aqueduct.*

There was the serious possibility that the expanding ice could smash the cast iron plates of the aqueduct so an attempt was made to break up and remove it – without a great deal of effect. The aqueduct survived this cold winter as it had many others in its long life.

Elsewhere the restoration ground to a snail's pace as the intense cold shattered steel tools when they were used, prevented concrete being made and brought diesel-powered engines to a halt.

Overleaf, top: *The cold was so intense that this photograph of the barge lock was taken from the middle of the river on foot! The picture shows the concrete and stone footbridge that replaced the earlier tramway swing bridge. This bridge had to be replaced by a new one at the lock tail to reopen navigation onto the river.*

Overleaf, bottom: *The river was so frozen that people walked, skated, cycled and even drove cars on it A large bonfire was even lit.*

On 2 March 1963, the thaw started and work resumed at its usual frenetic pace. It was here at Lock 52 that the original canal terminus was planned, later the plan was amended to have an arm of the canal beside Birmingham Road.

The cottage at Lock 40 was restored again in 1993.

The basin itself contained vast quantities of mud; after some difficult negotiations it was all used to cap over the borough tip nearby.

Even beneath Bridgefoot Bridge was a mess, the original towpath on the right had been broken down to prevent access from the Bancroft Gradens onto the rest of the canal.

The Herculean task of reopening the southern section of the Stratford Canal was more or less complete by the summer of 1964 and the rural nature of this beautiful waterway once again reappeared.

A flood of boats made their way down the canal for the reopening ceremony.

The theatre allowed the boats to moor against its gardens when the local council banned them from using the basin.

The canal was open again.

Open for business once more, but with very different cargoes.

Seven

Reincarnation

On 11 July 1964 the National Trust found itself the proud proprietor of thirteen miles of reopened canal. During the work the British Transport Commission had been thoroughly reorganised, with the Waterways section turned into a separate body now called the British Waterways Board, although civil servants and bureaucrats still far outnumbered actual workers. It's new constitution called for it to be more friendly towards the waterways and their users, the idea of boating for leisure was at last encompassed within its remit. Other canals were starting to attract their own societies and pressure groups. Co-ordinated by the Inland Waterways Association, the recreational use of the canal system was at last beginning to play a significant role. The winter of 1962 had frozen out the last few commercial users so that the future of the canals lay in the hands of private boaters and British Waterways.

The National Trust had to decide whether or not to continue with its involvement with the canal almost immediately; the five year lease was due to expire shortly after the reopening. After a detailed study they decided that they would exercise their option to purchase the freehold of the canal. Their study had concluded that it could be managed with two full time members of staff and a manager, with extra help in the form of prisoner work parties and volunteers. Several assumptions went into the planning of the canal's future which were likely to prove problematical. The first of which was that the canal had been restored; it had not. The canal's reopening had shown the extent of the work still necessary; many of the repairs had been on an ad hoc basis and would require further work. The projections of usage were guesswork and assumed the canal would remain a quaint little cul de sac, certainly not part of the Avon Ring with a couple of hire boat bases sending novice boaters bouncing off the locks. Perhaps crucially they had not fully appreciated the work that their canal manager, David Hutchings, had put into the project, and how difficult it would be to replicate. David had given notice that he intended to withdraw from active involvement with the canal shortly after it had reopened, he was exhausted from the three-year ordeal.

The new manager was Major Grundy, a stalwart of the IWA. Luckily the amount of traffic on the canal was low, and so the continuing process of restoration made some progress.

Through the mid-1960s about 300 boats a year made the journey down from Lapworth. Each boat paying a toll for the trip. At the time British Waterways were abandoning their toll system in favour of a time limited licence. The National Trust found that the income generated by their toll system was just about adequate to cover the running costs of the waterway. The IWA organised work parties to help improve the locks and towpath. The towpath in particular was in a dreadful state and, where the mud had been dredged onto it, rank vegetation grew in profusion, the path became a slithery mess each time it rained, and the mounds and ruts were a virtually insuperable obstacle to anyone wishing to explore the canal by foot. The canal was definitely regaining its old ramshackle charm, but it wasn't easily available to anyone without a boat.

As the 1960s drew to a close, leisure boating started to gain popularity across the whole country. British Waterways had been forced to provide a more secure business environment for those wishing to start hire boat businesses, and the result was a steady growth in the number of boats using all of the canal system. Some of the more intrepid crews pottered down the South Stratford. The result was a pleasing increase in tolls for the National Trust, and a massive increase in wear and tear which promptly absorbed all the money and more. Novice boaters contributed to some of the problems. Experienced crews were more used to canals maintained to better standards and continued to use working techniques that assumed there was someone to mend the canal after they had used it.

The National Trust started to wonder about the wisdom of running a waterway, not only was there the problem of the increasing deficit, they were getting embroiled in various disputes about their management style. The inland waterways scene was nothing if not fractious, various interest groups vying for pre-eminence or a particular policy. Robert Aickman had frequently despaired of getting all these disparate interests to work together to save the waterways system, and in retrospect it is nothing short of a miracle that he succeeded. The National Trust did not want to be involved in the incessant squabbles and started to contemplate a dignified

The newly opened canal soon found itself heading back to it's overgrown state. Access onto the canal was difficult even in the town centre.

Without adequate resources the National Trust had to allow the canal to gradually revert to its derelict state, The cottage at Lock 54 was demolished despite local protest. Plans to rebuild it have not found favour with the new owners, British Waterways.

withdrawal. Its experience with other waterways was equally difficult, the problem being in most cases that the people who originally proposed restoration felt as though they had a personal interest and tended to form cliques which discouraged new boaters. The Trust, dedicated to saving historic artefacts for the whole nation, wanted no part in these disputes.

The canal itself was falling apart under the increased usage. During its working life the boats had always been hauled by horses, however all the modern boats used motors, creating a damaging wash that eroded the banks rapidly. The very last thing that the canal needed was a massive increase in usage; which is exactly what happened when the Upper Avon Navigation opened in 1974. Suddenly the South Stratford Canal was part of a nationally renowned cruising ring. From a few hundred boats a year there were over 3,000, the result was catastrophic on the canal's structure. The National Trust's management system could not rely on volunteer labour to cope with the trail of broken paddles, boats sunk in locks, and an increasing barrage of criticism. The working deficit of the canal was in excess of £10,000 per annum, drawing funds away from projects that the Trust felt were more important to the nation's heritage.

The dry summer of 1976 caused the canal to be closed. Although this caused a drop in tolls for through passages, both Anglo Welsh and Western Cruisers moved their fleets down to the basin in Stratford and continued to run them from there. It was a scene reminiscent of the old working days, dozens of boats moored side by side across the basin each weekend, cars full of holidaymakers loading their luggage, generators and pumps chugging away. As a proper working scene, Stratford Council was not impressed. With no water coming down the canal, a 6in-diese-powered pump recycled water from the river back into the basin. Its noise, twenty-four hours a day, aggravated the situation and the National Trust received the back lash. Coupled with the official complaints, the canal system, including the Stratford Canal, was starting to attract a different type of user, vagrant boaters wound about the system, usually on dilapidated boats, bringing back the old fears of those on the bank. The National Trust was finding itself in between a rock and a hard place.

By 1977 the Trust announced that they were hoping to transfer the canal to a suitable trust, with something like the Avon Navigation Trusts as the model in mind. However, they received no credible proposals, and found themselves in the dubious position of contemplating closing the canal that they had worked so assiduously to save. The only option open was to start negotiations with British Waterways. The BWB classed the canal as a remainder waterway, and would give little in the way of guarantees as to any continued navigation unless the Trust was to bring the canal up to a decent standard. This was plainly out of the Trust's ability and negotiations dragged interminably onwards.

A deal was eventually struck that would allow the National Trust to finish its involvement with the canal and assure at least some sort of future. The BWB agreed to take over the canal if the Trust also handed over a sum of money to cover some of the maintenance backlog. There was no alternative, and in 1988 BWB gained control of the waterway and quite a few hundred thousand pounds.

The canal had almost regained its derelict character when BWB took over. The Trust had replaced many of the oak lock gates with a hideous modern steel design, but the towpath and offside banks were overgrown, the main channel was silted up so badly that in many places deep-draughted boats couldn't pass each other. Some valuable work had been carried out by the Waterways Recovery Group, but the navigation was in a sorry state. The BWB started to gradually replace the gates as they failed, embarked on a spot dredging program and towpath clearance. At long last Stratford Council was persuaded to take a far more positive attitude to the canal, to the extent of entering into a partnership with BWB to improve the towpaths within the town's boundary.

The new management style took many of the canal users by surprise. With national resources BWB could replace gates and paddle starts with their own design. The remaining cottages on

Work by volunteers from the local canal society and Inland Waterways Association help keep the jungle at bay, but it wasn't enough.

the canal were sold off as being irrelevant to the navigation. Projects to rebuild the ones that had been demolished were discouraged. Wherever possible the canal was improved to a standard that would have staggered even the original builders. The standard was that of an industrial canal, the new gates were built to a design suited to the Black Country, characteristic wooden paddle starts replaced with welded steel ones; the effect was similar to putting Formica onto a Chippendale table. The canal was being stripped of its identity and its beauty. The new management continued with this process by ensuring that virtually all the town's boat owners were deprived of their local moorings to make way for a time share business.

At long last the role of the canal in the economy of the area was acknowledged by the local authorities. 130,000 people boating their way to Stratford brought substantial revenue to the town each year and the district council decided to keep the canal basin in good navigational condition by sub-contracting British Waterways to dredge it, and providing a free site to dump the spoil. Improvements to the more visible areas of the canal, those close to the town centre, have been able to attract funding both from the local authority, British Waterways and an assortment of other sources. The canal's status as a remainder waterway continues, despite the National Trust's request for it to be made a Cruiseway. British Waterways have none the less embarked on a massive dredging programme which will ensure the canal's survival, but the gradual erosion of the canal's unique architectural features continues unabated. The Great Western Railway features such as mile posts and stop plank racks, so evocative of the story of the canal, are being ripped out almost faster than they can be catalogued.

Quite what the future holds remains shrouded in mystery. The canal that inspired Temple Thurston and Robert Aickman with its aura of tranquillity; the canal that inspired the restoration of the whole canal system, lies in the hands of a nationalised bureaucracy. Hopefully they will be able to find the resources and will to continue the restoration in a sympathetic way.

Vandalism was a constant problem, particularly in Kings Norton and the city outskirts. This is Lifford Lane Bridge, looking north to the junction with the Worcester & Birmingham Canal. Until the stop lock was built beneath the bridge, wide barges could use the canal as far as Hockley Heath.

Bridge Cottage, Mr Farr, the boat builder, had his yard here, the entrance to the dry dock is now covered by new stonework on the right.

British Waterways brought in new management and expertise, whilst retaining the canal's original staff.

Repair work was carried out where needed, but with non-traditional materials such as sheet steel piling. It seems rather like putting a plastic laminate onto a fine Georgian table.

Reopening the 1812 lock to the Grand Union Canal, which helped to conserve water.

The new lock and replica bridge are built with modern materials to maintain a veneer of the canal's heritage, but overall the canal is losing its unique character.

Appendix

Shareholders at the Inaugural Meeting

Name	No. of shares
Earl of Abergaveney	10
Richard Allen, Grocer of Stratford	10
John Allen of London	5
Samuel Aylworth of Halford	10
Fernando Babbington of Hockley House, Innholder	2
Elizabeth Barren, Widow of Stratford	3
Elizabeth Barke, Widow of Stratford	3
William Barke, Innholder of Stratford	5
John Barnhurst, Gent of Stratford	5
Phoebe Ball	5
Thomas Beach, Birmingham Steelmaker	3
John Berry, Birmingham Steelmaker	3
Sarah Belamy, Widow of Hockley Heath	1
Samuel Baylis of Henley	2
Thomas Biddle of Tamworth	1
William Bishop of Long Compton	5
John Boote, Gent of Atherstone	10
Alexander Boote, Gent of Atherstone	5
Francis Brookes, Yeoman of Aston Cantlow	1
Fernando Bullock of Berkshire	10
John Burman, Gent of Lighthall	5
Richard Burman, Gent of Tanworth	5
Thomas Burman, Gent of Tanworth	5
Jonas Barborough, Yeoman of Stratford	2
Richard Baker, Gent of Stratford	10
William Bond, Yeoman of Lapworth	5
Earl of Yarmouth	10
William Chambers, Esquire of Stratford	2
Francis Charles, Gent of Henley	3
John Checketts, Gent of Wilmcote	5
Thomas Conway of Birmingham	2
Samuel Cook, Yeoman of Lapworth	3
Clement Catterell, Gent of Moseley	2
Dean Corbett of London	5
Micheal Corgan of Chipping Norton	5

Issac Court, Watchmaker of Henley	2
Issac Court, in the interest of the poor of Henley	2
Thomas Cresswell, Gent of Stratford	10
Edward Bolton of Ham Hall, Staffs	10
Duke of Dorset	10
Ann Dennison, Spinster of Birmingham	5
Court Dewes, Esquire of Wellesbourne	10
James Davenport, D. of D., Stratford	10
Thomas Dobbs, Miller of Kings Norton	5
James Dolphin, Gent of Kenilworth	3
David Dolphin, Yeoman of Monkspath	2
John Drew, Gent of Pershore	5
William Eares, Esquire of Stratford	10
Richard East, Esquire of Stratford	10
Thomas Edkins, Miller of Welford	1
John Edwards, Cooper of Stratford	10
Philip Evans, Yeoman of Hockley	1
Revd James Eyre of Solihull	5
Thomas Featherstone of Packwood	10
William Field, Gent of Tanworth	5
Joseph Foster of Birmingham	3
Ashberry Fuller of Birmingham	5
Revd John Fullerton of Stratford	10
Thomas Gem of Birmingham	5
Revd Daniel Gacher of Wootton Wawen	10
Henry Geast, executors of, B'ham	5
Anthony Gibbs of Alne, Gent	3
John Gough of London	5
Richard Gough of Hampshire	3
Revd John Granville of Derbyshire	10
Marquis of Hertford	10
Lucy Handy of London	2
John Harding of Solihull	10
Judd Harding, Surgeon of Solihull	10
William Harding, Esquire of Hampton Lucy	10
William Hawkes of Birmingham	3
Peter Holford of Wootton Hall	10
Richard Hill of Kineton	10
Thomas Reeve Hobbins, Grocer of Stratford	10
Joseh Hykes, Yeoman of Kings Norton	1
Job Hobbins, Gent of Shottery	5
Henry Howard of Northumberland	10
Thomas Holmes of Beoley	10
William Hicks of Birmingham	5
William Hunt of Lincoln Inn London	10
Samuel Oliver Hunt, Gent of Stratford	10
Revd John Hunt of Welford	10
Thomas Hunt, Gent of Tanworth	10
Thomas Hunt, Gent of Stratford	10
Henry Hunt of Tanworth	10
William Hutton, Papermaker of Birmingham	2
Thomas Hooper, M.D. of Worcester	10

Jervoise Clerk Jervoise, Esquire of Hants	10
William James, Gent of Henley in Arden	5
William James, Innholder of Henley	1
Thomas Jenkins, Yeoman of Saltley	3
Charles Jenkins, Grocer of Stratford	5
Samuel Jarvis, Mercer of Stratford	10
William Johnson of Cambridge	10
Charles Ingram, Cooper of Stratford	10
Issac Ingram of Nuthurst	2
William Ingram of Nuthurst	1
James Ingram, Yeoman of Preston Bagot	1
Richard Jesson of West Bromwich	10
Jonathon Izod, Plumber of Stratford	10
James Jones of Birmingham	2
Samuel Johnson, Gent of Stratford	5
Revd Townsend Kendrick of Atherstone	10
Robert Knight of Barrels, Warwickshire	5
William Lygen of Worcester	5
Revd John Lucy of Charlecote	10
John Lane of Birmingham	2
Joseph Lavender of Evesham	5
Richard Lea, Innholder of Kings Norton	5
Henry Greswold Lewis of Malvern Hall	5
Mary Lloyd, Widow of Birmingham	5
Revd William Loggin of Long Marston	10
George Lloyd of Welcomb	10
Haneage Legge of London	10
Sir John Mordant of Walton	10
Charles Mordant of Walton	10
William Merriot	5
Richard Morland of Springfield House	5
John Merry of Aylsbury House	3
William Mills of Middlesex	10
John Mills of Billesley	10
Thomas Mills, Surgeon of Stratford	5
Robert More of Shelsley	1
Joshua Morgan of Tunbridge Wells	5
Charles Musgrave of Berkshire	10
William G Morris, Gent of Stratford	5
Margret Nanfan, Widow of Sherbourne	3
John Nott of Bath	5
William Oldaker, Mealsman of Stratford	10
Earl Plymouth	10
John Payton, Gent of Stratford	10
Newsham Peers, Esquire of Alveston	10
William Pulmer, Yeoman of Preston Bagot	1
William Penn, Gent of Stratford	5
William Penn, Gent of Red Hill	5
George Perrot of Fladbury	5
Andrew Perrot of Fladbury	5
John Perrot of Fladbury	5
Thomas Perrot of Fladbury	5

Charles Pestell, Surgeon of Stratford	10
Ann Porter, Widow of Birmingham	5
John Pountney, Yeoman of Kings Norton	5
Benjamin Parker of Birmingham	5
Richard Reeve of Beaudesert	10
Issac Pratt of Worcester	5
William Roberts of Bewdley	10
Wilson Aylsbury of London	10
William Russel of Birmingham	5
Thomas Nutton of London	5
Sir Edward Smythe of Wootton Hall	10
Thomas Sabin, Yeoman of Preston Bagot	1
William Salmon of Birmingham	2
Thomas Sheldon, Grocer of Stratford	10
Evelyn Shirley of Ettington	10
Ann Scheverall, executors of *	2
Smauel Smith, executors of *	10
Mary Snape of Birmingham	5
William Smith, Yeoman of Kings Norton	2
William Southam, Perukemaker of Stratford	5
Walter Stubs of Shropshire	5
Thomas Smith, Gent of Stratford	10
Sir John Throckmorton of Berkshire	10
Thomas Taylor, Joiner of Stratford	10
John Tarlton of Botley	5
John Tilbury of London	10
William Tompkins, Mercer of Stratford	10
Gore Townsend of Honnington	10
Mary Twamley, Widow of Sutton Colefield	4
William Twamley of Sutton Colefield	5
Samuel Twamley of Sutton Colefield	5
James West of Snitterfield	10
Theopholus Wolford of London	5
John Waring, Yeoman of Tanworth	3
John Wall of West Bromwich	10
Margaret Webb, Widow of Sherbourne	5
Thomas Wilmore of Birmingham	2
William Walford of Banbury	5
Revd John Whitmore of Stratford	5
Robert Wheler of Stratford	5
Robert Wheler (jnr) of Stratford	2
John Whittington, Yeoman of Preston Bagot	1
Edmund Wigley of London	10
Elizabeth Woolaston, Widow of Loxley	1
Thomas Miker of Culworth	3
Thomas Cook, Yeoman of Kings Norton	5
William Jordan, Yeoman of Kings Norton	5
Borough of Stratford-upon-Avon	5
Revd John Williams of Wellesbourne	10
Susannah Whyley of Staffordshire	2

* These shares were bought by the administrators of the deceased's estate.

Bibliography

Documents in the care of the Public Records Office, Kew:

Under heading RAIL 875
Minutes of the General Assembly of the Stratford-upon-Avon Canal Company
Minutes of the Stratford-upon-Avon Canal Company Board of Works
Miscellaneous documents relevant to the company

Under heading RAIL 886
Minutes of the General Assembly of the Worcester & Birmingham Canal Company

Records of Stratford & Moreton Tramway and various railways.

Documents in Shakespeare Birthplace Trust Records Office:

Stratford-upon-Avon Boat Register
Stratford Canal Papers
Stratford Tramway Papers

Midland Bank Archives

Published Sources

Waterways to Stratford	Charles Hadfield & John Norris (David & Charles, 1962)
Hadfield's British Canals	Charles Hadfield (Edited by Joseph Boughley) (Sutton, 1998)
Save the Stratford Canal!	Guy Johnson (David & Charles, 1983)
Stratford-upon-Avon, Portrait of a Town	Nicholas Fog (Phillimore, 1986)
The Borough of Stratford-upon-Avon	Minute Books
The Flower of Gloster	Ernest Temple Thurston (Williams & Norgate, 1911)
Shakespeare's Railways	John Boynton ([Publisher unknown], 1994)
The Two James's and the Two Stephensons	E.M.S.P. (David & Charles, 1961)
The Canal Duke	Hugh Malet (David & Charles, 1961)
Race Against Time	David Bolton (Methuen, 1961)